W9-AWZ-175

Creative Concrete Ornaments *for the* Garden

Making Pots, Planters, Birdbaths, Sculpture & More

Creative Concrete Ornaments *for the* Garden

Making Pots, Planters, Birdbaths, Sculpture & More

Sherri Warner Hunter

LARK BOOKS

A Division of Sterling Publishing Co., Inc.
New York

EDITOR
Jane LaFerla

ART DIRECTOR
Stacey Budge

PHOTOGRAPHERS
Evan Bracken
Stewart O'Shields
John Widman, pages 53, 54, and 84

ILLUSTRATOR
Olivier Rollin

COVER DESIGNER
Barbara Zaretsky

ASSISTANT EDITORS
Nathalie Mornu
Rebecca Guthrie

ASSOCIATE ART DIRECTORS
Lance Wille
Shannon Yokeley

PRODUCTION ASSISTANCE
Kristi Pfeffer
Shannon Yokeley

ART INTERNS
Chris Dollar
Brad Armstrong

EDITORIAL ASSISTANCE
Delores Gosnell

Published by Lark Books, A Division of
Sterling Publishing Co., Inc.
387 Park Avenue South, New York, N.Y. 10016

Distributed in Canada by Sterling Publishing,
c/o Canadian Manda Group, 165 Dufferin Street
Toronto, Ontario, Canada M6K 3H6

Distributed in the U.K. by Guild of Master Craftsman Publications Ltd., Castle Place, 166 High Street, Lewes, East Sussex, England BN7 1XU
Tel: (+ 44) 1273 477374, Fax: (+ 44) 1273 478606, e-mail: pubs@thegmcgroup.com, Web: www.gmcpublications.com

Distributed in Australia by Capricorn Link (Australia) Pty Ltd.,
P.O. Box 704, Windsor, NSW 2756 Australia

Manufactured in the U.S.A.

ISBN 1-57990-585-4

Contents

Introduction

Concrete is not hard! Well, it is and it isn't. Certainly skilled engineering is required for constructing modern-day concrete structures, and additives, mixtures, and finishing options are more sophisticated than ever, and then there's the surface...that's hard. But what's not hard about concrete is its accessibility, both in acquiring the raw materials you need and in finding information on tools and techniques for getting started. Most importantly, you'll find that it's certainly not hard getting hooked on working with this versatile and wonderful product.

Since writing my first book about concrete, I have continued to play, learn, create, and meet wonderful folks who have been willing to share their work and experiences with this material. With concrete at the core, the range of artistic work, garden solutions, and personal and public environments have exploded in recent years. Best of all, more and more everyday people, not just artists, are finding they can express themselves using concrete.

Whether you're an experienced artist, a person in search of a creative niche, or a gardener who wants that perfect pot, I would encourage you to read through these pages and open yourself up to concrete possibilities. You'll find the basic information you need for getting started, as well as more advanced techniques that will give you professional results. If you're familiar with my first book, you'll find new information on surface treatments, including coloring concrete and polishing, and more recipes for different concrete mixes, including lightweight and versatile hypertufa.

All the projects have been designed to make your garden even more beautiful than it is now. Imagine having a textured, exposed-aggregate bench tucked in a shady corner of your garden. A colorful polished garden table is the perfect focal point for serving a sun-dappled lunch or cocktails at sunset. There are mosaic butterflies for your butterfly garden, a whimsical snail plaque, vegetable steppingstones, and a bubbling fountain that will soothe you with its relaxing sound.

I'm fond of telling people, "Make your dreams concrete." Images of folk environments from around the world provided by Willem Volkersz are inspiring examples of how people have done just that by translating their artistic visions into concrete reality. All the designers and artists I've worked with to make this book possible have inspired me and I hope their work will inspire you too. Just remember, as you're mixing that first batch—concrete is not hard!

CHAPTER ONE

Concrete Basics

It's pretty amazing that three simple ingredients, cement, water, and aggregate, have had such a lasting effect on our lives and on the world's infrastructures, architecture, design, and creative expression. When mixed together, they make concrete, the basis for many of the buildings and objects we encounter daily. This chapter will give you some of the basic principles of working with concrete. Once you see how easy it is, you can start having fun exploring the creative potential of this incredible material.

Concrete Ingredients

Learning about concrete's ingredients and how they work together is important to getting the best results for your projects. Here's a quick summary of the concrete process start to finish. The cement and water make a paste that forms molecules on the surface of the aggregate, starting a chemical process known as hydration. During hydration the concrete hardens or cures. Curing is not the same as drying; in fact it's quite the opposite. When concrete is undergoing hydration, it needs to be kept damp and cool to allow it to cure to maximum strength. If new concrete is left to dry in the air or sun, the resulting concrete will be weak and probably have some cracking.

Cement

You'll be using the most common type of cement, Portland cement Type I, for the projects in this book. Portland cement is available in five different types, indicated by Roman numerals I to V. The difference between them is in their set and cure times, which relate primarily to construction specifications. In some cases you'll find a combination of numerals and letters to designate a blended cement with multiple qualities.

TYPE I This is the most common type of general-purpose cement and is readily available from building supply and home improvement centers.

A Brief History of Concrete

Egyptians used a mortar made from gypsum and lime to construct the pyramids. The Greeks continued to make improvements, but it was the Romans who made the most progress. By combining slaked lime—lime that had been broken down in water—with volcanic ash, they created a mix that most closely resembles our modern cement. During that time, similar construction advances were being made in Central America and Mexico.

When the Roman Empire fell, developments in cement were lost to that part of the world for centuries. In 1824 Joseph Aspdin, an Englishman who was a bricklayer and mason, patented the manufacture of hydraulic cement mixed from specific portions of limestone and clay that was fired in a kiln, and then ground to a fine powder. He named his product Portland cement because the color of the hardened material resembled the limestone quarried on the British Isle of Portland.

Portland cement was manufactured for the first time in the United States in 1871. The American Inventor Thomas Edison furthered kiln technology in 1902, leading to increased concrete production and popularity. Edison realized the versatility of concrete and even used it to make cast furniture. However his dream of mass producing low cost homes with a patented concrete casting method was never realized in his lifetime.

White and gray Portland cement in powder form (out of the bag) with cast concrete samples made from each

TYPE II It sets slower than Type I and generates less heat. This cement is often used for walkways that are subjected to de-icing chemicals during winter.

TYPE III This cement is referred to as high-early-cure, achieving most of its strength within the first week of curing. This makes it good for projects that need to have forms removed quickly, or when an area needs to be used shortly after construction.

TYPE IV A slow curing variety that generates very little heat from hydration. It's used exclusively for large masses of concrete, such as dams.

TYPE V This cement is formulated for concrete that will be in contact with heavy alkaline soil or groundwater.

Initials after the type number provide additional information, telling you what additives are in the cement.

Cement is usually packaged in water-resistant bags in a weight of 94 lbs (43 kg). When I buy a bag, I purchase it from a source that turns over their materials on a regular basis. You don't want to get a bag that has been sitting around a long time, or even worse, one that has a hole in it. As soon as a bag of cement is opened it begins attracting moisture and can get as hard as a rock. You may already have bags of cement at home that you're thinking of using for your projects, but if they feel hard or lumpy, don't use them. Keep in mind that cement is cheap and that the real value of your work is the thought and time you put into it. Using the highest quality materials you can always ensures the best results.

Portland cement is typically gray—the shades of gray vary according to where the base material was mined

Aggregates, clockwise from top center: river rock, perlite, peat moss, masonry sand, construction sand, crushed rock, vermiculite, pea gravel (center)

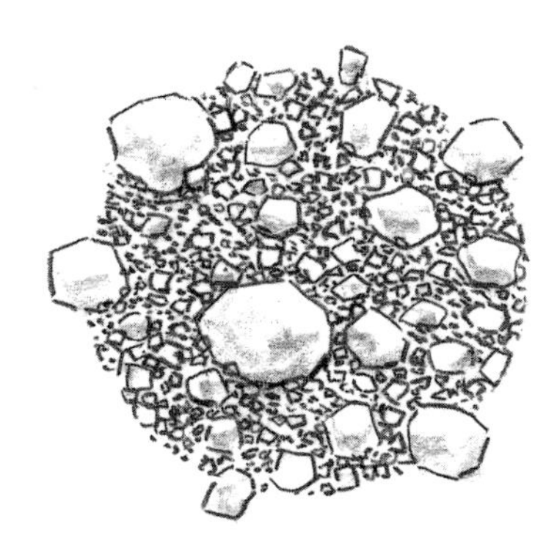

FIGURE 1

and processed. Portland cement also comes in white, which is made from further refining to remove most of the iron and magnesium. This extra refining makes the white more expensive—about twice as much as the gray—and difficult to find. You're most likely to find white Portland through a concrete company or at a building supply yard. The strength, set, and cure qualities for the white are similar to gray. The advantage to using white Portland is when you're adding color to your mix-the white will always give you a purer color.

Aggregates

Besides adding strength to the concrete, aggregates determine the quality of the concrete and its working properties. Some common aggregates are crushed gravel or rock, pea gravel, marble chips, play sand, and masonry sand. Aggregates can also be lightweight materials such as vermiculite and perlite—used by builders as insulation and by gardeners for soil aeration and drainage—that are used to reduce the weight of the concrete while giving it an interesting stone-like texture. Aggregates come in different grades from coarse to extremely fine (see the chart on page 12). In the building trade, aggregates are rated by number according to size—the higher the number the finer the material. A well-graded mix of aggregate is made up of various sizes of crushed rock or gravel and sand.

Aggregates also act as fillers to bulk-up the concrete. This helps reduce the amount of cement you use and can save you money. While there's a wide range of materials that can be used as aggregates, there's an equally wide range in their cost. But generally, per cubic foot, aggregates cost less than cement. This cost difference may be one of the reasons why a mix with a higher cement ratio to aggregate is often referred to as *rich.*

The type of aggregate you choose depends on the project you're making. For small sculptural pieces that will have a smooth texture, choose a fine aggregate. Masonry sand is finer than construction sand as an aggregate. If you're casting a large, weight-bearing element, you want to add a well-graded mix of sharp, crushed gravel and sand to your mix. As shown in figure 1, the smaller gravel shifts to fill in the spaces between the larger ones, while the sand and the cement particles, known as the *fines,* fill in the voids. This creates a strong matrix of materials that, when cured properly, will create a rock-solid mass.

Aggregate Chart

COARSE TO MEDIUM: STONE

Type of Aggregate	Crushed gravel/rock	Marble chips	Pea gravel	River rock
Where to Find	Building supply, home improvement center	Building supply, garden center, marble manufacturer	Building supply, home improvement center	Building supply, home improvement center, river beds
Qualities	Gray, angular	Most often white to gray, luminous	Natural, rounded	Natural brown tones, rounded

FINE: SILICA/SAND

Type of Aggregate	Masonry sand	Play sand	Marble sand
Where to Find	Building supply	Home improvement center	Building supply
Qualities	Sharp, light in color, usually ordered in bulk	Available in bags	Fine, white sand, also called swimming-pool mix

EXTREMELY FINE

See Pozzolans on page 13

MISCELLANEOUS

Type of Aggregate	Vermiculite	Perlite	Semi-precious stone	Crushed shells
Where to Find	Garden center	Garden center	Special order	Manufacturer, nature
Qualities	Natural material that is heated to expand, resembles mica	Natural occurring volcanic glass, also heated to expand, whitish in color	Wide range of colors	Textural, colorful

Concrete's strength comes from the hydration process as the cement and water molecules form on the surfaces of the aggregate. Flat, sharp surfaces like those found in crushed rock are considered to make the strongest bond. Always make sure that the aggregate's surface is clean of dirt, debris, and plant or other organic materials before you mix it with the cement. Never use beach sand. The salt on the sand will have an adverse reaction in the mix, creating an inferior product.

Water

Always use water that is clean and drinkable. Don't use muddy, oily, or toxic water. Never use salt water in your mix. If you do, the concrete will start out strong but weaken over time. The addition of salt also attracts moisture to the surface, increasing *efflorescence,* which is the migration of minerals that shows up as white deposits on the cured surface. If you use any steel for reinforcing the concrete, salt will increase corrosion of the metal and lead to a weakened structure.

Admixtures

Admixtures, also known as admixes or additives, are any materials added to a mix other than the cement, water, and aggregate. Admixtures are used to change the workability and strength of the concrete or to give the concrete different properties. For example, the most common admixtures used are air-entraining agents (see page 13)

Admixes, clockwise from top left: acrylic polymer, air entrainer (with bubbles), concrete bonding adhesive, water reducer (also called plasticizers or superplasticizers), latex polymer, polypropylene fibers, metal fibers (steel wool)

that make the concrete more workable and help it withstand the shock of freeze/thaw cycles. Admixtures can be liquids or solids. There are synthetic fibers that can be added to a mix to increase the strength of the concrete, while metal fibers can add color and interest. A description of some of the most common admixes follows.

AIR ENTRAINERS These soap-like substances produce a large amount of microscopic bubbles that form tiny voids in the concrete. This helps the concrete with expansion and contraction during the freeze/thaw cycle. Concrete with air entrainers should only be machine mixed to provide the needed agitation to create the bubbles. If a mix contains air entrainers, the initial after the cement type will be "A."

WATER REDUCERS These are sometimes referred to as plasticizers or superplasticizers and may allow for as much as a 15 percent reduction in the required water content. They also increase the concrete's strength and its bond to steel reinforcing rod (rebar).

ACCELERANTS Calcium chloride is probably the best known additive to speed up the set time of your concrete. It may be helpful to use when you have a short amount of time or are working in cool temperatures. Accelerants will weaken the mix over time.

SET RETARDER This is used to slow down the set time of your concrete and may be helpful when working in hot conditions.

LATEX POLYMERS These additives are usually a milky white liquid that is added to the concrete and can be diluted with water. They also act as a plasticizer so you can use less water in your mix and make the concrete more waterproof while increasing the strength. These polymers come from an adhesive base, which is why they're also used as a bonding agent, helping new concrete to adhere to older concrete surface. Acrylic polymers have basically the same properties and will provide good results.

POZZOLANS Another category of materials that can be added to your mix are pozzolans. They are silicates that combine with calcium hydroxide, a by-product of the hydration process. Pozzolans can also be used as extremely fine aggregates. Some pozzolans are available in blended cements. When added, they can improve your mix by improving the workability and by making the finished piece stronger and more water resistant. If a mix contains pozzolans, the initials after the cement type will be "IP." The most common pozzolans are silica fume, a by-product in the production of silicon metal; metakaolin, a kaolin based clay product; fly ash, a fine-grained glassy powder that is recycled from coal burning power plants; and blast furnace slag that is the ground-up residue from smelting metallic ore.

Premixed Concrete

You can buy a variety of concrete mixes that are premixed in a sack. All you need to do is add water and stir (so to speak). There are several different brands of these products, and you can easily find them at home improvement centers. They're usually available in 40-, 60- or 80-pound (18.2, 27.2, or 36. 3 kg) bags. The biggest advantage to them is that everything comes in one bag so you don't have to worry about storing quantities of separate ingredients. The flip side is that you're working with a predetermined mix without the opportunity to customize it.

If you walk down the aisle that has the concrete at your local home improvement center, you'll see a huge variety of premixes. However, only some of them will be recommended as an alternative to the custom recipes listed in the project instructions. (See pages 28 to 31 for custom recipes.) Usually premixes have rocks in them that are way too big for the techniques taught in this book. The three main premixes you should be aware of are the professional masonry mix, sand mix, and concrete mix.

Ready Mix

Ready mix is the term used for concrete that arrives in a truck, ready to use. There may be a point in time when you will need to explore the qualities and quantities of ready mix for large projects, but you won't find them in this book.

Tools and Equipment

One of the great things about working with concrete is that you don't need many tools to get started—basically a container to mix in and a pair of gloves will get you going. You may find you already have many of the tools you'll need. There are also many common items found around the home or workshop that you can use for fabricating molds or for finishing techniques. A few simple safety items will protect you as you work. As you get more involved with your projects, you may want to add a few specialized tools to your workshop.

Mixing Tools

For most of the projects in this book, you'll be mixing smaller amounts of material by hand. I always use a plastic tub with rounded corners in a size that accommodates the quantity I want to mix. Home improvement and building centers carry two or three sizes of mortar boxes. They are black plastic tubs with rounded corners that are convenient to use and inexpensive to buy.

Typically, you place them on the ground and mix the concrete in them using a mortar hoe. It looks similar to a garden hoe except its blade has two holes in it that allow the concrete mixture to move through them as you push the materials back and forth. You can use a regular garden hoe for mixing, but the holes in the mortar hoe mix the concrete more thoroughly. If you have a wheelbarrow, you can use it to mix and transport your concrete—just mix in one area, move the concrete to your site, and then move it to your clean-up area.

Mortar box, mortar hoe, and shovels

Plastic dish tubs and containers used for measuring and mixing

I find that the smaller mortar box makes a good container when hand mixing on a tabletop. Plastic tubs normally used for washing dishes can be used as well. A collection of smaller plastic containers can be useful when measuring and adding ingredients and water. The advantage to working with plastic is that it's strong, lightweight, and easy to clean.

Power mixers are a necessity for mixing larger quantities of concrete, or for the successful incorporation of some admixtures, such as air entraining agents (see page 13)—but you don't need to rush out and get one. The two main types used by the artists featured in this book are the cement mixer and the mortar mixer. The difference between the two is that a cement mixer's mixing blades are attached to the sides of the drum, and the drum rotates to mix the materials. A mortar mixer has a stationary drum with blades that rotate. While you can mix mortar in both machines, the mortar mixer is designed to thoroughly mix the smaller particles of sand and cement. If you use a mortar mixer to mix concrete containing larger aggregates, it may damage the blades.

Screw-lid, 5-gallon (19-L) mixer

Hoe and finishing tools, clockwise from top right: short-handled mortar (mud) hoe, small steel trowel, steel trowel, pool trowel, edger, margin trowels, pointing trowels

Another mixer on the market is about the size of a 5-gallon (19 L) bucket (see photo left). It has blades on the inside and a screw lid. Once your dry materials are in the container, you just screw on the lid and roll it on the ground to mix. Then add your water and roll some more. If you have any children around, they'll love to help you mix concrete with this method. I've had great results with this mixer.

A good option for mixing small amounts, particularly when you want to add an air-entraining agent, is to use a mixing attachment designed to fit on you power drill. You can get a well-integrated mix with very little effort.

Hand and Finishing Tools

A trip to a concrete distributor or the isle of your local home improvement store will reveal many more tools than the ones I've listed here. The following are just a few of the common ones that I'll refer to in some of the processes I'll be explaining.

SCREED This isn't a fancy tool you buy. It's the name used to refer to a straight edge (most often a board) used for striking off or leveling a poured or cast form. The process is called screeding.

MASON'S (OR BRICK) TROWEL AND POINTING TROWEL These are the tools we most often associate with bricklayers. They are typically diamond shaped and are used to deliver or place concrete to a specific area.

MARGIN TROWEL Similar in use as the pointing trowel but is rectangular in shape.

STEEL TROWEL While this tool is called a trowel, it's actually rectangular and is used in finishing concrete to make a smooth, polished surface. I sometimes use this tool for other than its intended purpose, and instead use it as a *hawk* or a tool that holds a quantity of concrete while working with other tools.

RUB BRICK This is an abrasive brick with a handle used to knock off and grind down bumps or smooth edges.

Sherri Warner Hunter, *Soul Searching,* 2003. 26½ x 18 x 19 inches (67.3 x 45.7 x 48.3 cm). Figure carved in polystyrene foam and covered in pigmented polymer fortified concrete; the base was cast in a polystyrene waste mold and then covered in a mosaic of ceramic, glass, and found objects. Photo by Gary Layda

Woodworking tools, clockwise from top right: miter box and backsaw, tape measure, angle, trim saw, duplex nails (two-headed), landscape nails, galvanized screws, drill bits, counter-sink bit, power drill

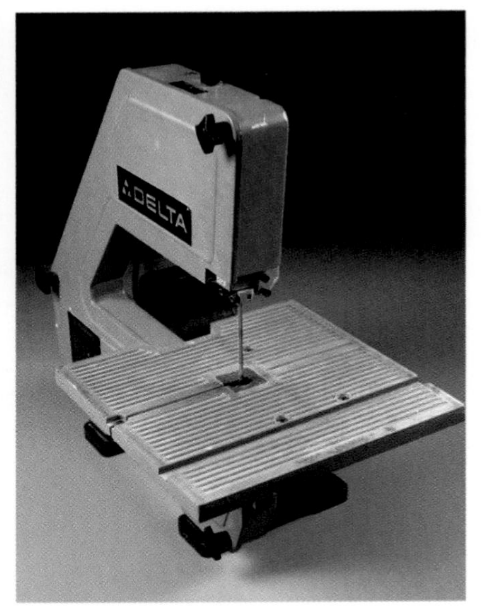

Band saw

Woodworking Tools

Several of the projects in this book use wood for building a mold or a form. Your basic hand tools—hammer, saw, and screwdriver—will come in handy, as will a tape measure and a square. Of course if you have power tools, things will go even easier. A power saw of any type, ranging from a circular saw to a table saw, band saw, or jigsaw, will help to cut boards more accurately. Some home improvement stores will cut boards for you but there will usually be a small charge for the service. Here are some other woodworking tools you'll need.

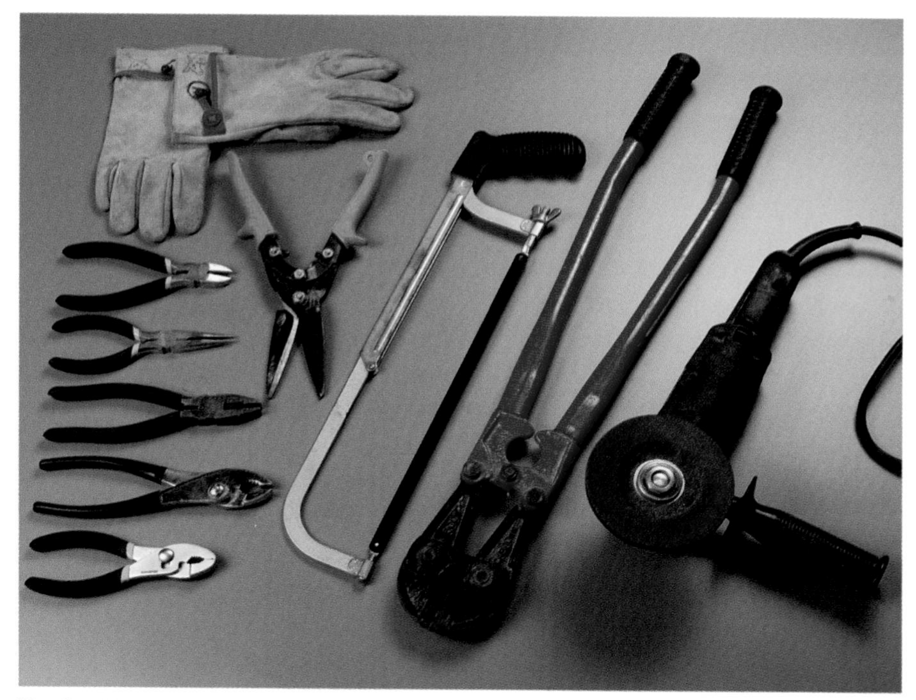

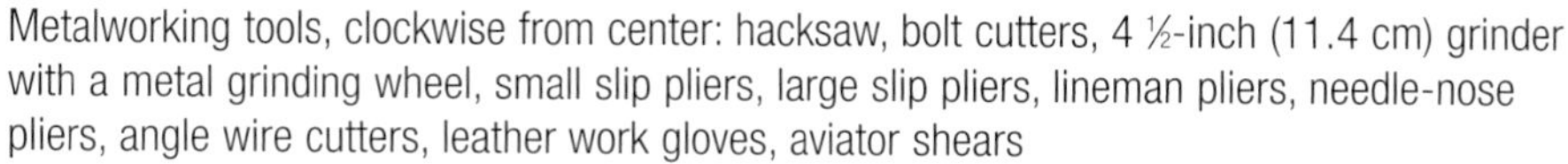

Metalworking tools, clockwise from center: hacksaw, bolt cutters, 4 ½-inch (11.4 cm) grinder with a metal grinding wheel, small slip pliers, large slip pliers, lineman pliers, needle-nose pliers, angle wire cutters, leather work gloves, aviator shears

Metal cutoff saw

FASTENERS It's always helpful to have an assortment of fasteners. You may already have drywall screws, either galvanized or regular, galvanized roofing nails, and common nails on hand. Another nail that is useful for building forms is the duplex nail. It has two heads—one that will snug to your board when hammered in, and the second head ½ inch (1.3 cm) further, that is used to pull the nail out when finished.

POWER DRILL If you are just starting to collect power tools for your projects, a good power drill is your best investment. You can get a variety of attachments that will help you with grinding, sanding, and mixing, in addition to drilling and screwing.

Metalworking Tools

Many of the armatures and reinforcing materials used in this book are made of metal. A good hacksaw, power cutoff saw, or a 4½-inch (11.4 cm) grinder with metal grinding wheels that can also be used as a metal-cutting tool, will come in handy. Following are some other tools you'll want to have on hand.

ANGLE CUTTERS These cutters are for cutting wire clean and close.

AVIATOR SHEARS Aviator shears differ from tin snips in that they have a built-in spring action that helps when cutting sheet metal and metal meshes used for reinforcing.

LINEMAN'S PLIERS These are strong-jawed pliers that will help you get a good grip when tightening wires. Each pair has a built-in wire cutter.

NEEDLE-NOSE PLIERS Their slender nose, as the name implies, lets you reach into tight areas. They usually have a wire cutting area as well.

BOLT CUTTERS Available in different sizes, they will help you cut through heavier metal rods. The larger sizes can cut through ¼-inch (6 mm) rod or #3 rebar.

Sherri Warner Hunter, *SAS Totem Series,* 2001. St. Andrews Sewanee School, Sewanee, TN. 6 to 8 feet (1.8 x 2.4 m) tall by 10 to 25 inches (25.4 x 63.5 cm) wide. Totem on left: *School Spirit.* Student-built project with elements carved from polystyrene and covered with polymer fortified concrete and alkali resistant fiberglass mesh; the dog "Coltrane" is concrete over a rebar and metal mesh armature and finished with embedded glass. Photo by the artist

Setting Up Your Workspace

There are no hard-and-fast rules as to where your workspace should be or how it should look. A carport or indoor workshop work equally well. But regardless of where or how big your workspace is, there are some basic considerations you'll want to make.

Do you have easy access to your materials for mixing?

Do you have an available water source for mixing and clean up?

Is your work area protected from extreme weather conditions during the construction and curing process? Too much wind, sun, and rain can affect the cure and make for weak concrete in the finished product.

Can you position your project so you can work on it without straining? Is it at a comfortable height?

Over the years, I've developed ways of working with concrete that are convenient and easy. Now, they're second nature to me. Eventually you'll develop your own shortcuts. For now, let me share my methods with you. As you read over them, you might get ideas for setting up your own workspace and how best to plan your work.

Setting Up

The size of a project determines if I work on a table or the floor. I generally work on a sturdy worktable for small and medium pieces, and on the floor for large projects. For larger medium-sized sculptures, I sometimes create a platform using milk crates with a sturdy piece of plywood as my work surface. The important thing is to find the height that is the most comfortable for you to do the work.

Whenever I set up, the first thing I reach for is sheet plastic (4 ml) to cover my work surface—it not only protects the surface, but also makes cleaning up much easier. When working at my table, I like to position my projects on a work board also covered with the 4 ml plastic. The work board makes it easy to move the piece to another location between work sessions.

Working in the Round

When I make a three-dimensional or sculptural piece, it's essential for me to see and develop it as a whole. Because I want the piece to be interesting from all sides, I need to constantly turn, move, and work around it. Using a work board helps me rotate the piece. I've also found that placing a smaller board under the work board makes it even easier to rotate. It allows me to turn the work board without twisting the plastic on my work surface. I also use a heavy-duty metal turntable or a sturdy lazy Susan—they're the ultimate solution for working in the round.

Clean Up

For clean up, I keep a couple of 5-gallon (19 L) buckets filled with water in my work area. I use the water to rinse my tools and hands while I'm working and to clean up everything else at the end of my work session. I also keep a supply of kitchen scrubbies and terry cloth rags around to help in the clean-up process.

You never want to rinse concrete in your sink. Concrete is a hydraulic material that will continue to set underwater—even in your pipes where it can ruin your plumbing. Clean everything you can with the water in the buckets.

To dispose of the water in the buckets, I let them settle overnight. The next day, I pour off the clear water, exposing the sludge that's settled at the bottom of the bucket. To get rid of the sludge, I scrape it out and put it in the trash.

Wheelbarrows and mixers are best cleaned outdoors with a blasting hose spray, sometimes aided by a stiff cleaning brush.

Large Installations

When you see images of larger concrete sculptures, walls, or personal environments (like those pictured on pages 115 through 117), you might be curious as to how they were made. Generally, they're built on site. You want to apply the same considerations mentioned for working with concrete to site-specific projects, but also consider other factors, such as the scale of the project, a working timetable, and probably even the change of seasons.

Many concrete environments are on-going projects that progress throughout the year as the weather permits, many taking years to complete. As an example, the Watts Towers in Los Angeles, California, created by Simon Rodia, took 35 years to complete (see page 116). Once finished, he gave the keys to the property to a neighbor and walked away from his beautiful construction never to return again.

Granted, that's on the eccentric end of the scale. What you need to know for now about working on larger projects is that, depending on the technique you use, you can develop them over extended periods of time.

Sherri Warner Hunter, *Rain Forest Treasures: Old World Bench,* 2002. Memphis/Shelby County Public Library, Memphis, TN. 3 x 15 x 2½ feet (.91 x 4.6 x .76 m). Carved polystyrene foam covered in polymer fortified concrete and ceramic and glass mosaic; reverse-cast pavers with pigmented concrete; pre-cast concrete walls have recessed area to accommodate inlay; this is part of a larger environment. Photo by Murray Riss

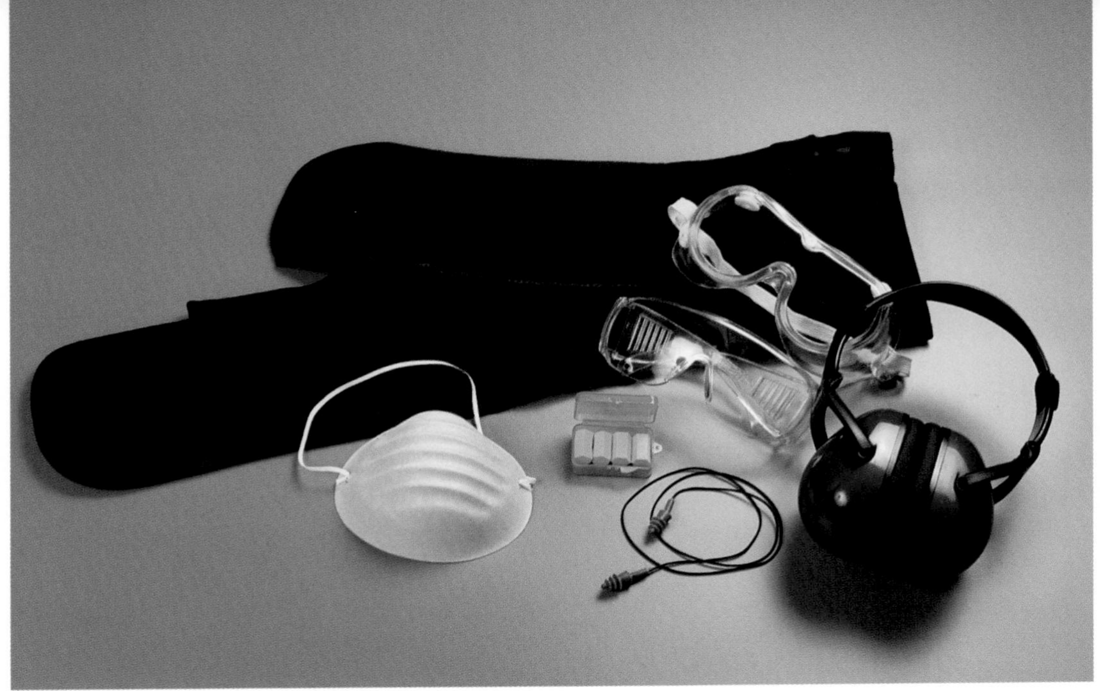

Safety equipment, clockwise from top: back brace, protective goggles, protective glasses, assorted ear protection, dust particle mask. (note: missing from this photo is a good respirator!)

Safety

Good safety practices prolong both your health and your enjoyment of working in any creative medium—concrete is no exception. One of the key components of concrete and many of its additives is silica. When inhaled over long periods of time, the particles can cause sinus and lung irritations leading to bronchitis if proper safety precautions are not followed. This is more common for miners and processors of cement, but is a factor you need to be aware of. By simply wearing a high-quality particle dust mask when you mix dry ingredients, grind, drill, or do anything else that creates dust from concrete or its components, you can protect yourself.

You want to protect your hands and eyes as you work. Having good work gloves is an easy safety precaution to take. They protect your hands from the caustic elements in the concrete and the abrasive nature of the aggregates. The cost and quality of your gloves may vary, but I've found that yellow, household cleaning gloves work well, are easy to find, and are inexpensive. I also keep a box of latex surgical-style gloves on hand when a finer touch is needed for more detailed work like modeling or embedding. Safety goggles should be worn at different stages of your project. It's a good habit to wear them when you work with any power tool, whether you're building a wooden mold or grinding concrete.

For those of you with weak backs, consider getting a back brace, they're the kind almost everyone working in home improvement centers wears. Also, use good lifting practices—lift with your legs, not your back. Better yet, consider working on projects with a friend so the two of you can share the joy, I mean, weight. You want to practice good safety habits, but mostly, use good common sense.

One additional item I'd like to recommend is a good hand cream. Concrete always draws out moisture, even from your skin. When you wash your hands after working with concrete they might feel slimy or soft, but don't be fooled. Later on they will begin to feel dry, very dry. Be good to your hands—use the hand cream.

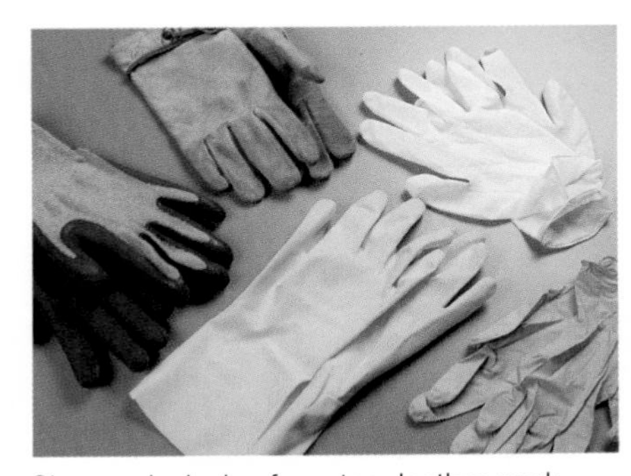

Gloves, clockwise from top: leather work gloves, latex surgical gloves, nitrile disposable gloves, latex cleaning gloves, rubber-coated knit gloves

Mixing Concrete

Whether you mix by hand or machine, the first thing you need to do is determine the quantity you'll use of each ingredient for your mix. All recipes for mixing concrete will refer to parts. A part is whatever you use for measuring to get the correct ratio of ingredients. Construction professionals will usually just use a shovel because they've developed a feel for the correct amount they need to use. Others will use a box or a bucket. For most of the quantities I mix, I use various sizes of plastic food containers—butter tubs and cottage cheese and yogurt containers. It doesn't really matter what you use as your part measure, just keep in mind that you must always use the same size for each ingredient. Also, avoid tightly packing ingredients when you fill your measuring container. If you scoop loosely, you have a better chance of getting more equal parts for each ingredient.

Ideally, you want to mix the correct amount of concrete for the project you're making. You don't want to have a lot left over, and you don't want to have to mix another batch just short of being finished. However, you'll find that both will happen to you at some point. As fun as concrete is, there's a lot of science and math behind it. Professionally, concrete is calculated in cubic amounts: cubic yards, cubic meters, cubic feet, and so on. To calculate the volume of a box-shaped form use the following formula: length x width x height = volume. To figure out the volume of a cylinder use the formula: pi (3.14) x radius squared x height = volume. Dividing the volume by 1728 will give you the amount needed in cubic feet. Elder Jones (see pages 151, 160, and 161), as an example, knows that he will get 3 cubic feet of concrete from a bag of cement. He consistently mixes three parts sand to one part cement. He also does the math and uses a measuring container that is equal to a cubic foot to determine just the right amount of material to mix.

You could also arrive at a close calculation by first filling your mold with the aggregate you're going to use. For example, if you're using a sand mix to cast a steppingstone, fill your steppingstone with sand, measure the amount of sand it took, and then calculate your portion of cement based on the sand measurement. The cement adds very little to the volume of the mix. The fine particles actually fit between the grains of sand once water is added.

The best idea is to keep some kind of record as you develop your mixes for each project. For instance, once you've figured out how much volume it takes to fill a steppingstone mold, make a notation of the container you used for measuring. Having that information will save you time and materials the next time you try to create that project. When I was in art school, students in the ceramic department always had their little black books—not for addresses—but for information on glazes, kiln-firing logs and schedules, preferred clay bodies, and such. A similar type of journal or sketchbook can help you

Sherri Warner Hunter, *SAS Totem Series,* 2001. St. Andrews Sewanee School, Sewanee, TN. 6 to 8 feet (1.8 x 2.4 m) tall by 10 to 25 inches (25.4 x 63.5 cm) wide. Totem on left: *School History.* Totem on right: *Nature.* Student-built project with elements carved from polystyrene and covered with polymer fortified concrete and alkali resistant fiberglass mesh. Photo by the artist

document your discoveries and duplicate your successes. You'll find that having a record of how long a recipe remains workable, how long it takes to set, or even a note on the day's humidity can be a valuable resource.

Once you know the parts you'll be using and have a general idea of the volume you're mixing, you can select an appropriate-sized mixing container. Try to mix a quantity that you can use in two hours or less, keeping in mind that the temperature and humidity will have an effect on how long you'll be able to work with your mixture. The drier and hotter the day, the less time you have to work.

Even though working with concrete can be very spontaneous, there's a certain amount of planning that will add to your ultimate success. Molds should be prepared and armatures standing ready before you even begin mixing the concrete. When using a mold or a form, mix a little extra to ensure that there's enough to top it off. Larger projects may require multiple batches to fill the form.

You can even plan ahead to use up leftover concrete by having smaller projects at hand and ready to complete. Several artists I've known have found creative ways to use excess concrete. Howard Finster used a variety of found objects and recycled containers as molds to cast elements that were later incorporated into a commanding mixed-media wall, and Elder Jones is constantly adding to his patio.

Mixing by hand

Mixing Concrete by Hand

Forget any images you have of concrete sliding out the back shoot of a rotating truck. Mixing concrete by hand or with a hoe works for most projects. Remember to wear your dust mask and gloves when measuring and mixing dry ingredients. Starting with the aggregate first, then the cement, measure your dry ingredients and place them in the container. The cement powder is very fine. If you put it in your container first, it will have a tendency to stick to the sides. Mix the dry ingredients together until they are uniform in color and texture.

Measure your dry ingredients starting with the sand (aggregate).

Mix the ingredients together thoroughly.

Next add your water. It's difficult to say exactly how much you'll need. Each recipe will have a recommended amount, but that amount could vary from batch to batch depending on the humidity, amount of moisture in your aggregate (sand), and even the temperature of your materials and tools. Start with about three-quarters of the recommended amount. Make a depression in the center of the dry ingredients and pour the water in it. Mix thoroughly, being careful to incorporate dried materials from the bottom and corners, until your batch is the right consistency, adding more water as needed. One of the keys to successful concrete work is to use as little water as you need to achieve the consistency needed for the process you are using. The "right" consistency will depend on what type of project you're making. You'll want a clay-like consistency for modeling and a muffin-batter consistency for casting a stepping-stone.

Mixing by Machine

When I need larger batches or want to use certain types of admixes, a mixer comes in handy. Air-entraining agents (see page 13) need the extra agitation and mixing capabilities of a power mixer. Poly fibers and color seem to disperse more evenly as well when mixed this way. When mixing by machine there's a difference in how you add the ingredients. First add about half of the recommended water, then your aggregate. Add the cement as the mixture is rotating. Finally, slowly add the balance of water until you have reached the right consistency. At a certain point, the moisture balance will quickly tip, so be very careful with your mixing. Make any notation on adjusted mixtures so you can duplicate them if you're working on a larger piece with multiple batches.

Begin adding water.

Add just enough water to get the proper consistency for the technique your're using.

Electric concrete mixer with water source and gravel located conveniently nearby

Working Stages

Once your concrete is mixed and you start to work with it, you'll notice that it will go through different stages or *set points.* During theses stages, the concrete has begun hydration. This transformation enables different processes to occur successfully and predictably at each set point (see figure 2). Once you are able to recognize the set points, you'll know what you can do with the concrete at each stage. Keep in mind that when certain set points pass, you've reached a point of no return and may not be able to finish or refine the material as you had hoped.

After about 30 minutes, the first or initial set will occur and the concrete becomes firmer. If the concrete gets firmer faster than that, it could actually be what is known as a false set, and the mixture can be brought back to a creamy working consistency by simply reworking the concrete with your hands to redistribute the moisture.

After two to four hours, concrete cast in a form or mold, such as a stepping-stone, will feel firm. The water that collected on the surface, known as *bleeding,* will have evaporated. This is a good stage for writing or incising into the concrete. If you're wet carving (see page 77), this is the point where you would remove the form and begin your piece. The concrete is very green at this point, meaning that it is very fragile and needs to be protected from any hard impacts. One of the true beauties of concrete is that it can be worked and transformed at every different stage of hardness. As the casting

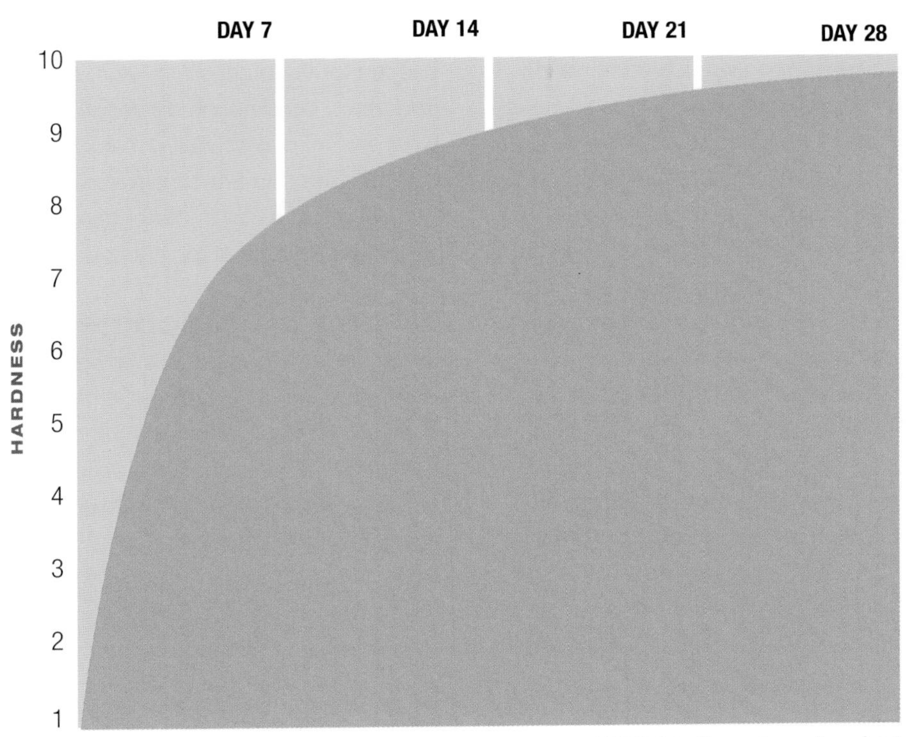

FIGURE 2. Concrete curing chart

continues to harden, you will need to adjust your carving tools and working techniques accordingly. Use your notebook to jot down which set points provide you the best opportunity for various forming techniques.

As you work, and your mixture continues to get firmer, you may find it difficult after a time to continue working at all. As a rule, you should never re-temper or add water to your mix. Sometimes I've found it necessary to bend that rule, particularly if I'm at the end of a session with just a little more work to do.

Adding a few spritzes of water from a spray bottle and a little remixing have worked for me, but again, it's not recommended.

Curing

After you have finished a project, the concrete needs to cure. In order for it to strengthen, concrete needs the necessary amount of moisture at the right temperature to allow the hydration process to take place. Remember, you want concrete to cure, not dry.

The first week of curing is the most important for fresh concrete. As you can see from figure 2, there is a dramatic curve showing the hardness that is achieved during this time. This is the time that you need to be misting your piece and/or keeping it covered. Concrete will continue to harden over its entire lifetime but is often considered fully cured after 28 days.

Concrete covered in plastic while curing

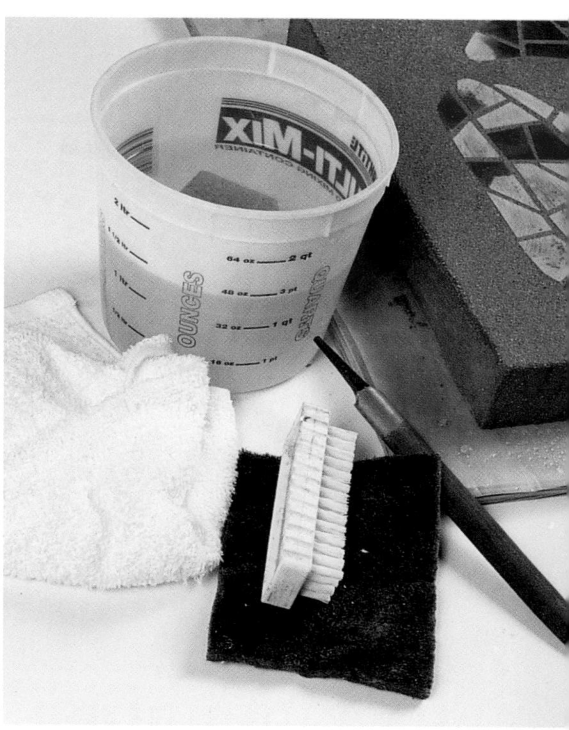

Tools for cleaning concrete, clockwise from top: water and sponge, file, kitchen scrubbie, stiff nailbrush, terry cloth rag

To ensure proper hydration and curing, mist your work at the end of a work session as needed, then wrap the piece in plastic. Large plastic trash bags or 2 ml plastic drop cloth work best for this, as shown in the photo above. Try to keep the concrete in a cool location out of direct sunlight or wind. Another method for curing is to wrap the piece in water-soaked newspapers, towels, or burlap. However, there may be some surface discoloration if you use this method. Smaller pieces, once they've reached the final set stage, can be submerged in a bucket of water to cure.

If you're working on site or on large pieces that are placed outside, covering or wrapping a piece may not be practical. In these instances, you may need to mist your work with a hose three or four times a day during the recommended curing times. Recommended cure times and any variations that are needed for specific techniques will be included in the project instructions.

Cleaning Concrete

At some point you will need to clean your concrete or remove residue from materials you are using in conjunction with the concrete, such as embedded materials used as surface treatments. The best habit to get into is to try to work as clean as you can, wiping off any embedded pieces with a damp sponge as you complete an area. This works better on some materials than others. Non-porous glass and glazed-surface ceramics are the easiest to clean (glaze actually being melted glass). Materials that are more porous, like unglazed ceramics, rocks, and shell can be more difficult to clean.

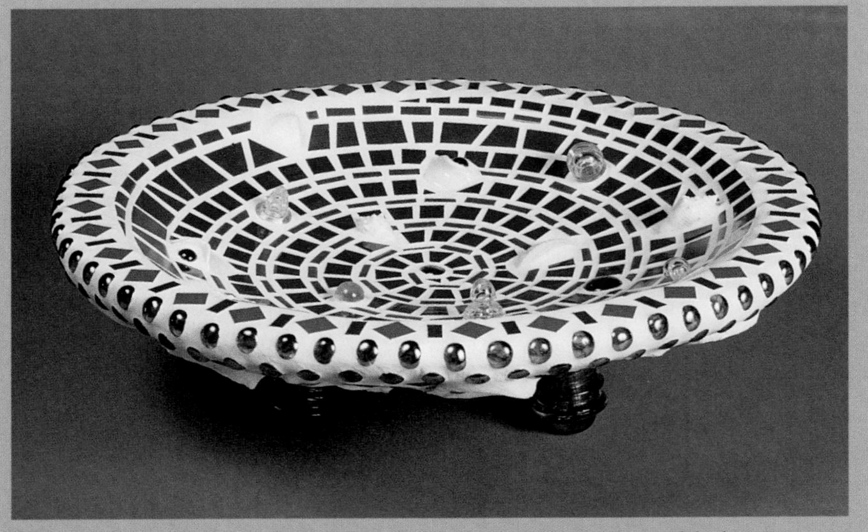

Sherri Warner Hunter, *Mediterranean Day Dreams.* 22 (diameter) x 6 inches (55.9 x 15.2 cm). Sand-cast bowl covered in ceramic, glass, and misc. mosaic; glass insulators serve as legs. Photo by John Lucas

Timing is a big factor in cleaning. You want to clean a piece after it's hard enough not to be damaged by the cleaning. But if you wait too long, the concrete will already be hardened, making cleaning more difficult. If you can return to work on an embedded piece within 12 to 24 hours, a stainless-steel wire brush, water, and elbow grease will remove almost any unwanted debris. Other handy cleaning tools are a stiff nailbrush, old toothbrushes, kitchen scrubbies, and a wooden craft stick. I use the wooden stick to chip off any small lumps without worrying about scratching my materials. After scrubbing or scraping the surfaces, I use clean water and rinse thoroughly—you don't want the residue to be allowed to dry back onto the surface.

Sometimes you need to clean the concrete itself because of mineral deposits, staining, or for surface preparation. There are some commercially produced solutions that are available for removing efflorescence, which are deposits of salts that leach from the concrete, usually white in color, that appear on the surfaces of masonry, stucco, grout, or concrete. Efflorescence is one of those unexplained occurrences, and there is no guaranteed way to avoid it. Most producers of cement-based materials will even include a disclaimer on the possible occurrence of efflorescence.

Concrete surfaces also need to be cleaned before the application of most colorants. Specific coloring products may suggest a cleaner manufactured by the same company, while warm water and a scrub brush will work adequately for others. Pressure washing may be necessary for larger pieces to remove debris or residue from casting techniques or to thoroughly clean before proceeding with color applications. Refer to Chapter 3 under Colored Concrete for more specific information on surface preparation for colored-concrete surfaces.

USING ACIDS FOR CLEANING

Another approach to cleaning concrete is the use of chemicals, specifically muriatic acid and sulfamic acid. Note the word acid—it's a logical tip-off to the level of caution you'll need when working with these products. Read any instructions on the packaging labels thoroughly before attempting to use these products. When I use them, I make sure to wear long pants, a long-sleeve shirt, closed toe shoes and socks, a mask, goggles, and gloves. I also put the animals inside and assure everyone that this is not a spectator activity. When I've finished using the acid, I pour the remaining solution into a 5-gallon (19 L) bucket of water to dilute it further before disposing of it safely. I then use the hose to rinse my work area, my tools, gloves, and goggles. Safety first!

MURIATIC ACID Muriatic acid is easily found at your home improvement center either in swimming pool supplies—it's commonly used to balance the ph of pool water—or in the paint department. You'll want to work outside with a garden hose nearby. Don't mix more than you need. It doesn't take much solution to clean an average-size piece, and you don't want to have lots of extra leftover to dispose of when you're finished.

With safety gear on, and using a plastic container for mixing, pour one to two cups (.24 or .48 L) of water into your container. To avoid splashing the acid, always pour the water in first and then add the muriatic acid. Don't lean directly over the bottle when you're opening it. As you open it, you'll notice steam-like vapors coming from the bottle. Do not inhale them. Mix a solution with about a one to five ratio: one part acid to five parts water. When you brush the solution onto the concrete surface, you will see it fizz—that's the acid at work. Too much acid will eat away at your surface so don't be tempted to make the mixture too hot, meaning adding more than the one to five ratio of acid to water.

I generally try to work from the top of the piece down by first going over the whole piece with a larger scrub brush and then working on individual pieces or smaller areas with a very stiff nylon brush or a wire brush. Rinsing regularly with water as you work flushes off the debris so it doesn't dry back onto the surface. The water also neutralizes the acid—the higher the percentage of water, the weaker the acid. When you've finished cleaning the surface embellishments, use the large scrub brush again to go over the entire surface of the piece, including the bottom and the interior. Then rinse thoroughly.

SULFAMIC ACID Sulfamic acid comes in dry crystal form and is usually found in the tile section of your home improvement center. Two to three ounces in a gallon of warm water will provide a strong enough solution to clean a light haze of residue. This chemical doesn't have the same hazardous fumes as muriatic acid but still needs to be handled and stored with extreme caution. When working with this acid, I use the same protective clothing and follow the same safety procedures I do when using muriatic acid.

Sherri Warner Hunter, *Rain Forest Treasures: Poison Dart Frog,* 2003. Memphis/Shelby County Public Library, Memphis, TN. 25 x 48 x 45 inches (65.5 x 122 x 114.4 cm). Carved polystyrene foam covered in polymer fortified concrete and ceramic and glass mosaic; part of a larger environment. Photo by Murrray Riss

It's All in the Mix

Fortunately there is a nice variety of premixed sacked concrete readily available. Many of these will be specified as one of your choices for making the projects, but as you continue your creative exploration in concrete, you will probably become interested in customizing your mixes. The following collection of recipes has come from a variety of sources. Many of the project artists have provided their personal favorites that they've developed by trial and error or taken from a magazine article some time back. Through your own experimentation, you might come up with a new mix that works perfectly for you. (Remember to keep that journal!)

The cement used for these recipes will be gray or white Type I Portland. The sand (silica) can be play sand, masonry sand, or construction sand. Note that each recipe has a number. They will correspond to the numbers given in the project instructions to indicate which recipe to use.

A reminder about water: you only want to add as much water as is needed for the mix to do what you need it to do. Each project will refer to the desired consistency since many of these recipes could work well for different techniques.

Start by gradually adding the suggested amount of water, add more if needed, and stop when it feels right.

1. Basic Concrete Mix

This is the basic mix used for most utilitarian purposes like footing, pads, and foundations. It's ideal for permanent installations. The larger gravel provides an inexpensive bulk and added strength. Because the size of the gravel in the mix is ¼ to 1 inch (6 mm to 2.5 mm) you won't use this mix very often for projects.

1 part Portland cement

2 parts sand

3 parts gravel

Approximately 1 part water

Add just enough water so that all the particles are well coated with cement paste and the mix is not crumbly.

2. Decorative Aggregate

This mix is similar to the Basic Concrete Mix except the aggregate is ¼ inch (6 mm) or smaller. You can use an assortment of aggregates to create a decorative surface by washing off the top layer of concrete. You can see this decorative surface in residential projects, such as driveways and patios, but it has been adapted for some smaller projects in Chapter 4.

1 part Portland cement

2 parts sand

3 parts decorative gravel, such as pea gravel, river rock, or marble chips

Approximately 1 part water

3. Basic Sand Mix

This is a versatile mix that you can use for many projects. You can alter the consistency based on how you will be using the mix, adjusting your water amount carefully when you do this. This mix is good for casting steppingstones or for modeling.

1 part cement

3 parts sand

Approximately 1 part water

4. Fine Sand Mix

(from Elder Jones)

If you're working on a carving, you don't want a tool to catch on a small rock. This could scratch the surface or even leave a hole. By sifting the sand in this mix, you're ensured of eliminating any larger aggregates before they can become a problem. Most of the time when working with this mix, you want to add enough water to end up with a consistency that resembles muffin batter.

1 part cement

3 parts sifted sand*

Approximately 1 part water

*If substituting a dry premix for this recipe, sift the bagged ingredients before mixing.

5. Super Sand Mix

This is a good substitute for the Basic Sand Mix. However, since it uses air-entraining agents, you will need to use a power mixer.

1 part cement

2 to 3 parts sand

Fibers (polypropylene)

1 teaspoon (5 ml) air-entraining agent

Approximately 1 part water or polymer admix

This mix is best used as a first application. If the surface looks fuzzy, don't worry; it's caused by the poly fibers that can be easily burned off.

6. Mortar Mix

Even though you can use a sand mix as mortar, true mortar includes hydrated lime. This addition makes the mortar mix creamier in texture and easy to work with for masonry projects. You can also use this mix as a stucco coat, either in its natural color or by adding color to suit your needs. Like the Fine Sand Mix, aim for the muffin-batter consistency, which is stiff enough not to drip, but sticky enough to stick.

1 part cement

¼ to 1¼ parts hydrated lime

3 to 6 parts sand

Approximately 1 part water

7. Hypertufa Mix A

(from Elder Jones)

This mix uses peat moss and perlite to give the concrete the appearance of textured stone. The addition of polypropylene fibers helps reduce shrinkage cracks and adds tensile strength. If you're using a 1-gallon (3.8 L) container as your part measure, ⅓ cup (.1L) of fibers is recommended. You want to lightly pack the fibers in your measure, and to avoid forming any clumps, separate the fibers before adding them to the mix.

1 part cement

1½ parts peat moss*

1½ parts perlite*

Polypropylene fibers

Water

*Plain potting soil, without fertilizers, can be substituted for the peat moss/perlite combination.

8. Hypertufa Mix B

(from Virginia Bullman)

Virginia uses this hypertufa as a modeling mix. Working with metal armatures, she builds up layers to form her wonderful sculptures. In most works, the hyertufa is both a mosaic base and a finished surface.

1 part Portland cement

1 part sifted peat

1 part fine vermiculite

1 to 2 parts water

9. Hypertufa Mix C

(from Bill Loney)

1 part Portland cement

1 part sand

1 part (in whole or in combination) of peat moss, sphagnum moss, vermiculite or perlite

1 to 2 parts water

10. Hypertufa Mix D

(from Andrew Goss)

1 part Portland cement

1 part sand

2 parts peat moss

water or acrylic/latex additive

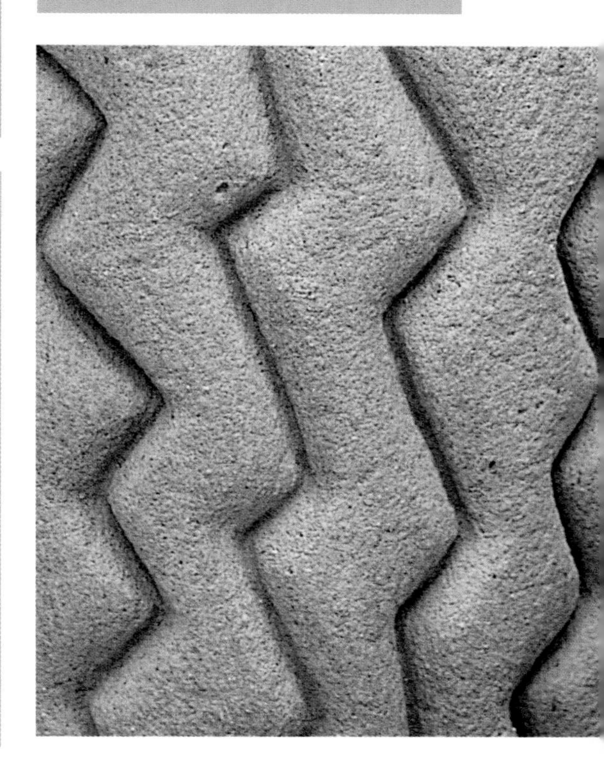

More About Hypertufa

The interest in hypertufa has been a growing one. That's evident from the artists' work included in this book. The development of this manmade material is the result of two interests, both related to gardening.

Hand-carved troughs and sinks that were used to water and feed farm animals have become gardening treasures. These rough-hewn antiques are ideal for planting miniature alpine and herb gardens but their availability is limited and they are quite heavy. Concrete planters are one option, but the weight factor is still there.

In the 1930s, it was discovered that alkaline-loving alpine plants, which also thrive with good drainage, grew very well in carved-out pockets of the Tufa rock, a spongy rock found near limestone deposits. Over the years, water would wash out some of the mineral components and deposit them to form this porous rock.

Hypertufa was developed to have some of those same qualities. It's lighter than a regular concrete or stone, and forms a strong but has a porous surface that resembles its namesake. This weatherproof material can be molded, cast, and carved. Turn to page 114 to see how you can even help your planters resemble ancient stone by encouraging the growth of moss on the surface. Miniature alpine plants, herbs, and Bonsai trees will love your handmade planter and troughs.

11. Vermiculite Carving Mix

To prevent the lightweight vermiculite from separating and floating to the top, slowly add the water to the dry ingredients. You want the mixture to be a creamy texture with the consistency of mayonnaise for casting your forms for carving.

1 part cement

2 parts vermiculite

Approximately 1 part water

Variation: Replace the cement and vermiculite with white Portland cement and perlite.

12. Polymer-Fortified Concrete

There are variations on polymer-fortified concrete recipes but this recipe is one that works very well for me and is what I use to create many of my concrete forms.

3 parts cement

3 parts sand

1 part acrylic polymer

¼ part water

13. Polymer Admix

You can make any mix stronger by adding latex or acrylic polymers. Combine the two ingredients, and then use them as a substitute for the plain water in your mix.

1 part polymer

4 parts water

14. Basic Fiber Cement

(from Lynn Olson)

There are no measurements for this recipe because it is mixed by feel. Start with a small amount of cement in a bowl. Mix in enough water to make a thin, watery consistency. Unroll a pad of steel wool, separate the fibers into thin strips, and add them to the mixture. You can cut the steel fibers with scissor to convenient lengths. Work them into the watery mixture until every fiber is coated with the cement. Add a little more cement to the mixture, pulling the fibers apart as you're mixing, to form a clay like consistency without any lumps.

Cement

Steel wool

Water

15. Fill Paste

(from K.C. Linn)

This is actually a recipe I adapted from K.C. Linn. She uses a variation of this to create her stamped surface textures, while I used these ingredients to back fill small holes that appeared in the table casting. See the polishing section on pages 92 to 94.

2 parts cement

Pigment (optional)

1 part bonding adhesive

Enough water to make a paste about the consistency of a thick cream

16. Super Fiber Cement

(from Lynn Olson)

Mixing the materials to a clay-like consistency makes this mix responsive to sculptural forming techniques. You can create your own ideal mix for use over armatures by experimenting with these ingredients.

White Portland cement

Metal fibers (steel, stainless steel, or brass)

Water and/or latex polymer (an emulsion that is 70% water)

Silica fume (or selected pozzolan)

Superplasticizer

See page 72 for more information on how to develop this mix.

17. Finish-Coat Fiber Cement

(from Lynn Olson)

Apply this mixture over the Fiber Cement. Folding in colored pigments after you've mixed the ingredients creates colored lines that will be accentuated in the final polishing.
Experimenting is the key to developing a successful mix.

White Portland cement

Carbon fibers

Latex polymer (an emulsion that is 70% water)

White metakaolin

Superplasticizers

18. Mesh Sculpture Concrete

(from Andrew Goss)

Andrew recommends combining your admixes with water to help with their distribution.

6 cups (1.4 liters) sand

4 cups (1 liter) Portland cement

⅔ cup (156 ml) stone dust (screened limestone or marble)

⅓ cup (78 ml) metakaolin

⅙ tsp. (.8 ml) air entrainer (optional)

1 tsp. (5 ml) superplasticizer (optional)

1-2 tbsp. (15 to 30 ml) plastic (poly) fibers

As small an amount of water as possible

19. Smooth Coating

(from Andrew Goss)

Use a smooth coating to rub into cracks or holes or to overcoat as a final smooth layer. If you leave out the stone dust, then make sure this coating is not too thick or you may get shrinkage cracks.

1⅓ cup (312 ml) Portland cement

⅔ cup (156 ml) stone dust (for very smooth, you can leave this out)

2½ tbsp (40 ml) metakaolin (if no stone dust, increase to 5 tbsp [80 ml])

As small an amount as possible of latex solution (1:1 latex to water) or acrylic solution

Slurry

When you're adding layers to concrete, you want the strongest bond possible. There are two things you can do: first, make sure the piece you're working on is damp before you begin layering, otherwise the concrete will pull moisture from the layering material; second, apply a binder known as cement slurry.

To make slurry, you mix cement with a liquid until it is the consistency of thick paint. When I mix a batch for a medium-sized project, such as a birdbath base, I use an 8-ounce (28 g) plastic container as my measuring unit and mix the slurry in a larger plastic food container. I find that this makes plenty for a good work session.

20. Basic Slurry Mix

1 to 2 parts cement

1 part water

¼ part water, no more

21. Fortified Slurry Mix

1 part cement

1 part polymer admix

¼ part water, no more

CHAPTER TWO

Forming and Construction Techniques

Where do you begin when you have an idea for a project? You might start by making a rough sketch or a small model that takes that idea to a design. Then as the design takes shape, you can choose the appropriate forming and construction techniques you'll need to use. For instance, if you've designed a square trough, you'll want to cast a square piece of concrete. Better yet, you might want to put an additional form inside the square so you won't have to carve out the center. Or, if you want to make an animal or figure, you'll probably need to make an armature. And, as the scale of the piece increases, you'll want to increase the gauge of materials used for the armature's construction.

There are a number of simple forming and construction techniques included in this chapter. They'll provide the reference you need to successfully complete the projects in Chapter 4. More importantly, they'll give you a good base of information for adapting your own designs and projects as you continue to explore the remarkable possibilities of concrete.

Simple Casting

Simply put, casting is pouring or compressing a material that is in a somewhat liquid state into a mold or confined shape. Once the material has become hard, the mold or shape is removed. You might also think of it this way: a mold is a negative shape. Once that shape is filled and allowed to harden, the results form a positive. For example, if your mold has a raised area on it, the resulting positive will have a depressed area.

Variations of casting techniques can be used for both artistic and functional projects including bowls, planters, decorative panels, benches, and, steppingstones. You may want to start with a small, basic casting technique to gain an understanding of the principles. This will allow you to experiment without investing too much in time, energy, and materials.

Molds, clockwise from top: assorted plastic bowls including takeout and storage containers, plastic tubs, constructed wooden mold with screed, metal puzzle-piece mold with reinforcing metal, metal frame with plastic liner, commercially available cardboard casting tube

Selecting a Mold

The word mold is used to reference a structure that will produce the same likeness each time. In construction, the word form is used to reference constructed structures that will produce a specific shape. If the structures are constructed the same way again, you will get the same likeness. But generally a form is constructed for a specific use, such as columns or stairs. In some instances, I will reference a form to use for a casting. I also use the word form as an indication of physical shape or sculptural appearance. Hopefully, the context in which I use the words will keep things clear.

For most of the projects in this book, I'm going to concentrate on simple mold making or on using items that you find around the house. You'll quickly realize that finding molds for simple castings is an easy task. A variety of materials can be used for molds and forms. Plastic, wood, metal, plaster, foam, and sand are all great mold-making materials. There are also more sophisticated materials available for making detailed castings or more complex forms.

Whatever material you choose for making your mold, there is one important consideration. How will you remove the hardened concrete from the mold? Some materials are flexible and can be peeled off a hardened casting. Other rigid mold materials, like plaster or polystyrene foam, can be chipped or broken away from the finished casting. Molds used only once and then destroyed in the unmolding process are referred to as a waste mold.

If you're making a mold of a sculpture that has undercuts, those sections that protrude and curve under like an ear on a human head, then you might make a rigid mold out of plaster in multiple pieces. You can then remove the pieces from the casting and reassemble them to create a piece mold that can be used for additional castings.

PLASTIC MOLDS

You can buy commercially produced steppingstone molds at many craft stores, home centers, or on-line sources. You can also find interesting molds if you go down the kitchen and home accessories aisles of your local discount shopping store. Start looking at bowls, containers, and toys through fresh eyes to imagine their casting possibilities. Even take-out food containers have great potential.

WOODEN MOLDS

If you want to cast a crisp angular shape, wood is a good choice for the mold. Small molds can be constructed by just screwing together a few boards. If you want your mold to be used for multiple castings, you might consider the construction example shown on page 129. If you want to make a wooden mold for a larger casting, then you'll need to reinforce the sides of your mold so it doesn't bow under the pressure of the large volume of concrete. Also, you want to make sure you've sufficiently secured the sides of the mold because you don't want to have what's known as a blowout. That's when the side of a mold or form comes loose while the concrete is being poured into it, causing the concrete to go everywhere. It's not a pretty sight.

METAL MOLDS

Metal can be used effectively for molds and forms. Elder Jones prepares simple molds using sheets of guttering metal or roof flashing (see the metal puzzle-piece mold on page 34). His puzzle-piece steppingstones (page 160) are constructed from 18-gauge steel. It's soft enough to bend, but will still hold its shape when filled with concrete. The same gauge of galvanized steel, available at your home improvement center, is also rigid enough to hold its shape during the casting. Roof flashing wouldn't be strong enough to hold its shape for this type of mold.

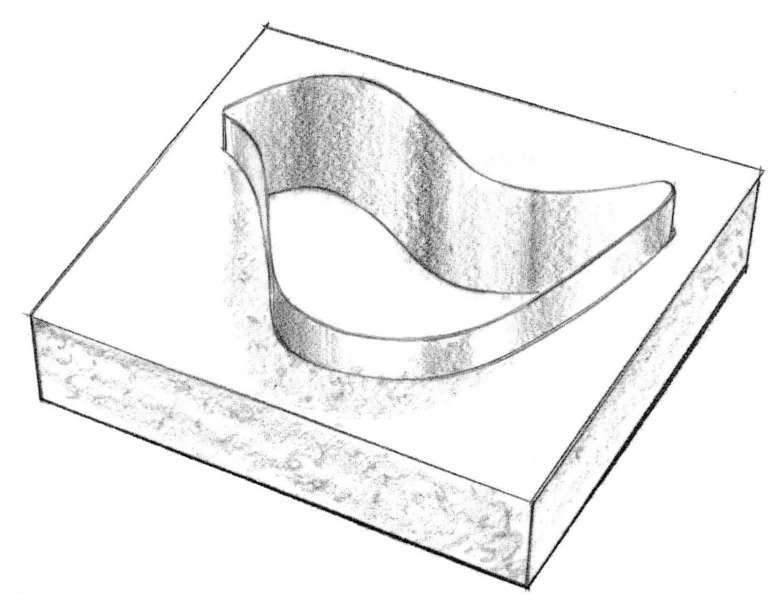

FIGURE 1

POLYSTYRENE WASTE MOLD

You make this mold by cutting a shape out of polystyrene foam and then filling it with concrete—it's that easy. Well, just about. It's also a good idea to wrap duct tape around the perimeter of your foam for added reinforcement, particularly if you've made your cut to the center from an outside edge. Sometimes I use roof flashing to line the negative shape in the piece of polystyrene to ensure smooth edges, as shown in figure 1. Once the concrete has hardened, the polystyrene can be broken away and the flashing, if used, peeled away. You can make more complex molds by cutting and stacking the foam. Insulation foam board can be used to create another exciting effect, the casting of a textured negative relief (see pages 45 to 47).

FLEXIBLE MOLD

Flexible molds make it possible for you to work out an idea or design in one material and then cast it in concrete one time or a dozen more. It's best to use a flexible-mold system, a pre-packaged set of compatible materials that are marketed together to make a durable, flexible mold for casting in concrete. One of the beauties of today's technology for this process is in the mixing. Most flexible-mold systems only require mixing equal parts, by volume, of the two components that create the material. Previously, most measuring had to be by weight, requiring a more elaborate work setup. Now you can just mix and go.

SAND

A bag of sand and some water—what could be easier? Sand is a wonderful, versatile mold-making material. Most of the molds I've mentioned so far require that you put the concrete into the shape. When you use sand as the mold, you'll generally put the concrete on top of it. A mounded shape might be just the right contour needed for the inside of a birdbath (see page 144). Or, a wooden frame with a layer of sand might help nestle bottles being cast in concrete for a glassy panel (see figure 1 on page 159).

Before you begin to form your shape, you need to add enough water to the sand so it holds its shape when you squeeze it in your hand. If your intentions are to reproduce a clean form, cover the shaped sand with plastic before applying the concrete. At other times, you might want to cast directly on the sand form to pick up some of the texture for a rustic surface.

Mold release agents, left to right: spray cooking oil, spray lubricant, petroleum jelly, motor oil, commercial mold release, dish soap

Mold Release Agents

Mold release agents are substances that are applied to the mold to insure that the concrete won't stick to the mold surfaces. A variety of release agents can be used; however, certain ones work better on certain surfaces. In the project instructions, you'll be told when you need to apply a mold release agent. I suggest you become familiar with your choices and use the one(s) that work best for you.

Commercially produced mold releases are available as sprays, brush-ons, or dry powders. A number of readily available products found around your home also work well, including motor oil, diesel fuel, petroleum jelly, spray lubricants, and non-flavored cooking sprays. Make sure your mold is clean before you prepare it for casting. You want to apply an even coat of your selected release agent to the entire surface until it looks shiny. As you apply the release agent, pay particular attention to corners and details. Avoid making drips, puddles, or large deposits of the agent.

I use motor oil, diesel fuel, or petroleum jelly on wooden molds because their dense and semi-water repellent nature prevents them from seeping into the naturally absorbent wood. I prefer using a paintbrush to apply the oil and fuel, but rub on the petroleum jelly. Plastic or metal molds are easily prepared with a spray lubricant or cooking oil. Some plastic molds develop surface scratches the more you use them. The surface scratches make the concrete grip the mold, and the more scratches, the tighter the grip. Even though a spray release agent might have worked fine when the mold was new, you may find it necessary to change to a heavier agent, such as petroleum jelly, as it becomes more scratched.

Plastic sheeting provides its own release. That's why I always work on plastic-covered surfaces. I also use a thin ml plastic to cover sand forms. It gives me a smooth surface when I want one, and the plastic also allows me to easily pull the casting away from the sand.

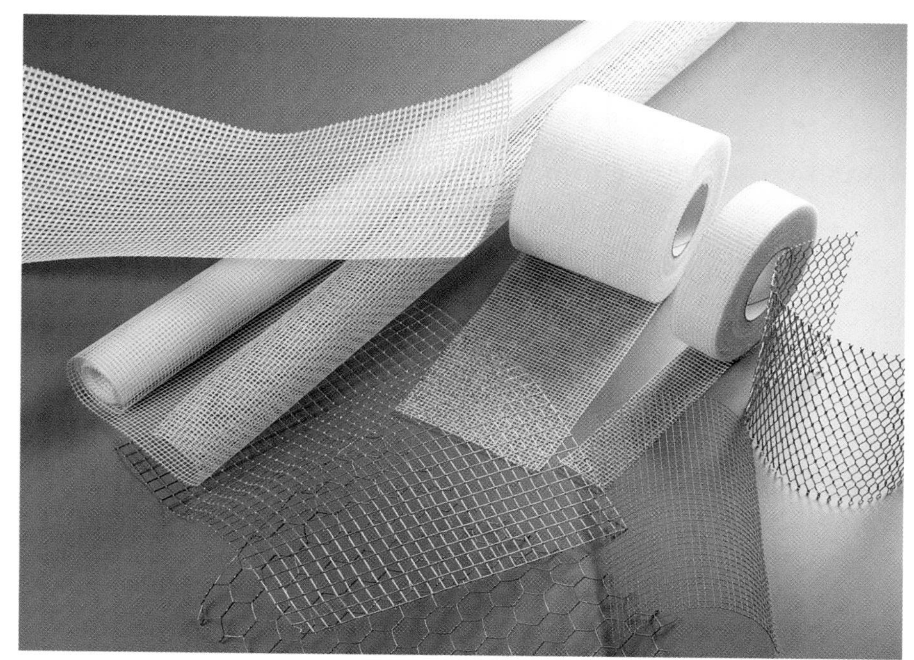

Reinforcing materials, left to right: alkali resistant fiberglass fabrics, self-adhesive AR mesh rolls, expanded metal mesh, chicken wire (front), hardware cloth

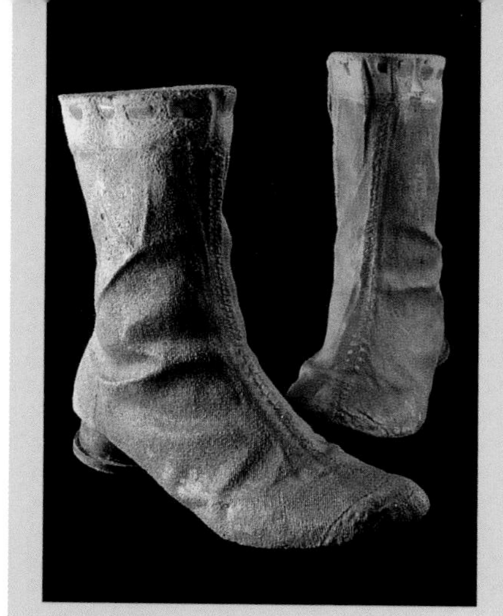

Kem Alexander, *Boots,* 2004.
8 x 3 inches (20.3 x 7.6 cm).
Old boots, metal squares; boot forms filled with concrete, surface cleaned with grinder and muriatic acid and water, sprayed with acrylic finish.
Photo by Art Image Studio

Reinforcing the Casting

Unless you're making very small objects, you'll want to include some type of reinforcing material in your mold during the casting process. The reinforcing material will add strength and integrity to the piece. If your concrete piece develops a crack, the reinforcing material should hold it together until you can make repairs.

The size of your piece will help you determine the type of material to use. A steppingstone might use a piece of ½ inch (1.3 cm) galvanized hardware cloth, while a slab being cast to use as a bench would benefit from ½ inch (1.3 cm) rebar. Another factor to consider when using various sizes of mesh is that you want the largest aggregate in your mix to easily pass through the spaces in the mesh. If they don't, you'll inadvertently create layers of aggregate by trapping the larger pieces on one side of the mesh.

Preparation is also important when using reinforcing materials. In many cases, the reinforcing material will be inserted midway through the casting process, so you want to have the materials measured and cut prior to mixing your concrete. Other projects may need to have the reinforcing materials wired together and suspended in the mold. If you're using a mesh, it's important to cut it smaller than the size of your mold. If the reinforcing material touches or protrudes into the sides of the mold, it will make it difficult for you to remove the finished casting, and may even cause it to break.

Fibers can also be added to your mix to help with reinforcing. Polypropylene or alkali resistant (AR) glass fibers can help inhibit cracking, improve water resistance and increase freeze/thaw durability. AR fiberglass fabrics provide a flexible reinforcing material that will contour to curvilinear or organic shapes.

Your preparation for these materials may require that you separate fibers or cut the fabric to desired shapes or sizes. Lynn Olson uses metal fibers, steel or brass wool, in his sculptures (see pages 72 and 172). Not only are they the basis for his fiber cement, they also add to the beautiful patinas on his finely finished surfaces.

Casting a Simple Mold or Form

As you prepare your work area, take into account the logistics of casting. Will it be easy to physically pour the concrete into the mold? Is the mold on a surface that you can move or turn if needed? If you're working outside, will the weather cooperate? Will you be working in a shaded area? Each of the casting projects in Chapter 4 will detail how to construct or find the mold needed. Once you've selected your mold, follow these instructions.

Materials and Tools

Concrete Mix: Premixed Sand/Topping Mix, Mortar Mix, Commercial-Grade Mason Mix, or Mix 2, Mix 3, Mix 4, or Mix 5...almost any mix you want depending on your project.

Mold

4 ml plastic to cover the work area and work board if using one

Mold release agent

Reinforcing material: ½inch (1.3 cm) hardware cloth

Aviator shears

Container to mix concrete

Screed

Paper or synthetic towels (optional)

Putty knife

Sponge

Water container

File (optional)

Instructions

1 Cover the work area and board with plastic. Construct or select your mold. Apply the selected mold-release lubricant to the inside of the mold (photo 1).

2 With the aviator shears, cut the hardware cloth ½ inch (1.3 cm) smaller than the inside of the mold and set aside.

3 Mix enough concrete to fill the mold. As you begin to cast, take your time, and make sure that all of the corners and edges are filled. Fill the mold halfway.

4 Position the hardware cloth in the mold so it doesn't touch the sides of the mold (photo 2). Continue to fill the mold until the concrete brims (photo 3).

5 To achieve a strong casting, you want to eliminate large air bubbles, and you do this through vibration. Small molds placed on a workbench can be vibrated by picking up a corner of the board and tapping it up and down against the table surface.

PHOTO 1

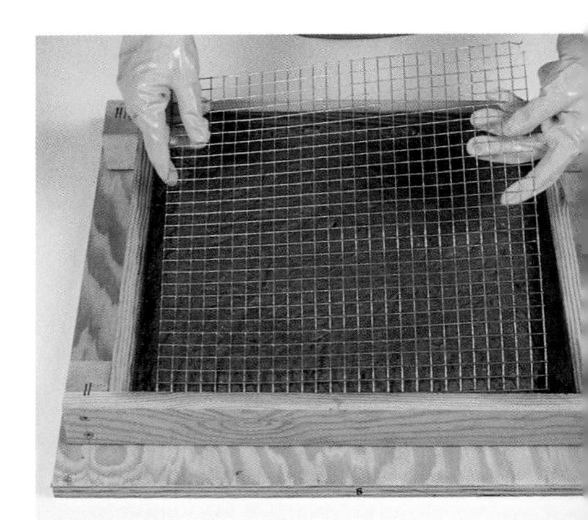

PHOTO 2

PHOTO 3

Marsha Hoge and Dixie Stephens, *Garden Edging,* 2003. 12 x 216 x 204 inches (30.5 x 548.6 x 518.2 cm). Basic sand mix, fresh elephant ear leaves. Photo by artists

Some molds may require tapping on the side with a hammer. A small electric sander held on the side of a mold can also create enough of a vibration to cause air bubbles to rise to the surface.

6 Screed or level off the top of the mold (photo 4). To do this effectively, position the straightedge board you're using as your screed across the back edge of the mold. Move it in a side-to-side sawing motion as you move the board toward you. Don't clean off your work board after screeding. This excess material around your mold actually works as a dam to stop additional concrete form seeping out of the mold. Later, you may even need a little of that concrete to finish off your piece.

7 Allow any bleeding, the water that has accumulated on the surface, to evaporate. You can also lay sheets of paper or synthetic towels on top of the casting to absorb some of the excess water. Repeat if more water collects.

8 Allow the casting to remain undisturbed for five to six hours if it's a small mold (say the size of a steppingstone), or until the concrete feels firm and no water appears when you rub you finger back and forth over the surface.

9 Clean your work board of excess concrete. If you've made a reverse-cast mosaic (like the example used here), reserve this material to finish your piece, and follow steps 13 and 16 for Reverse-Cast Mosaics on page 104. Clean off any concrete that might be on the edges of the mold by using a plastic putty knife or sponge (photo 5). This will help eliminate chipping when the mold is removed.

10 Remove the mold (photo 6). See Removing a Mold on page 40. Cover the casting with plastic, and leave undisturbed for 24 hours. The casting will be significantly harder after this period. File the edges of the casting if needed, and continue to let it cure.

PHOTO 4

PHOTO 5

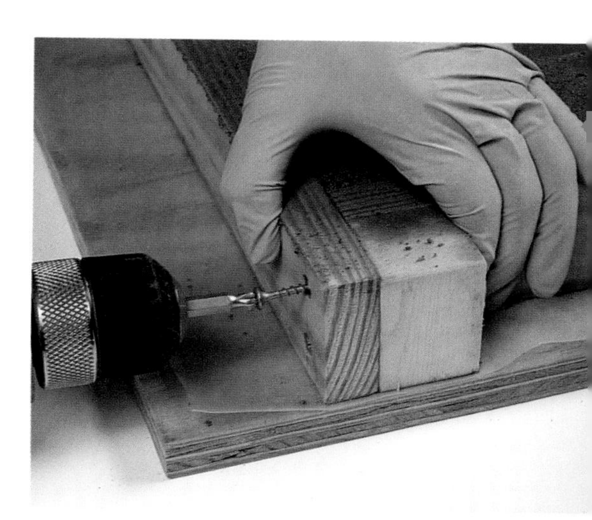

PHOTO 6

Removing a Mold

You want to wait for the right time to unmold your piece, and that will vary with each casting. The concrete needs to be firmly set. For instance, when casting a steppingstone, I generally unmold it after about six hours. The casting is firm but still workable. At this stage, I'm able to grout reverse-cast designs (see page 102) and smooth edges. But keep in mind that the concrete is also fragile at this stage, and any additional work needs to be done carefully. When you have a larger, more detailed mold, or are using a softer mix, such as hypertufa, you'll want the concrete to stay in the mold longer. Up to two days or longer is quite normal.

It's erroneous to think that your concrete will just slide out of the mold. You'll first need to remove any excess concrete from the edges of your mold. The concrete may want to hold onto the mold as you try to pull it away, which can cause chips in your casting. You want to gently work flexible molds until they're free from the cast surfaces and then push the piece out or pull the mold off. With assembled molds, remove any hardware, tape, clamps, or bindings that have held the mold together. Then carefully pull the sections away from the casting. Your casting is *green* at this stage, meaning it hasn't reached its full hardness, so work carefully. Once your casting is out of the mold, allow your piece to continue to cure by covering in plastic and keeping it damp.

Kem Alexander, *Dangerous Bowl*, 2003. 15 x 9 x 6 inches (38.1 x 22.9 x 15.2 cm). Concrete, cut nails; hand packed, nails inserted into setting concrete, cured piece cleaned with craft drill then placed into a bath of muriatic acid and water, then a bath of baking soda and water. Photo by Art Image Studio

Casting a Simple Bowl

You'll see there are several easy ways to create a bowl form. This simple bowl is made with the help of some plastic bowls from the local discount store. The resulting form can be used as a planter or an element for a fountain. Try to select bowls that are rounded, don't have a foot or bottom rim, and, if possible, a minimal top rim. They don't need to be the same shape, but one should be about 2 inches (5 cm) less in diameter than the other.

Materials and Tools

- Concrete Mix: Premixed Sand/Topping Mix, Mortar Mix, Commercial-Grade Mason Mix, Mix 3, Mix 4, Mix 5, or any of the Hypertufa Mixes
- 2 plastic bowls, 1 large 1 small
- Mold release agent
- Container for mixing concrete
- Small bag of sand (to fill smaller bowl)
- Putty knife
- Sponge
- Container to hold water
- Pliers

Instructions

1 Coat the inside of the larger bowl and the outside of the smaller bowl with a mold release agent (photo 7).

2 Mix the concrete to a clay-like consistency. Pack the concrete firmly into the bottom of the large bowl to a thickness of 1 inch (2.5 cm) or more (photo 8).

3 Center the small bowl on top of the concrete. Add the bag of sand—rocks can be substituted—(photo 9). This will help weight the small bowl so it will not float up as you add more concrete.

4 Continue to add concrete between the two bowls. Take care to tamp, or tap and compress, the concrete as you go to eliminate any air pockets (photo 10). Keep the smaller bowl centered as you continue to fill the sides (photo 11).

5 Trim off any excess concrete (photo 12) and smooth the top edge with a sponge (photo 13). Cover with plastic and let the bowls sit undisturbed over night.

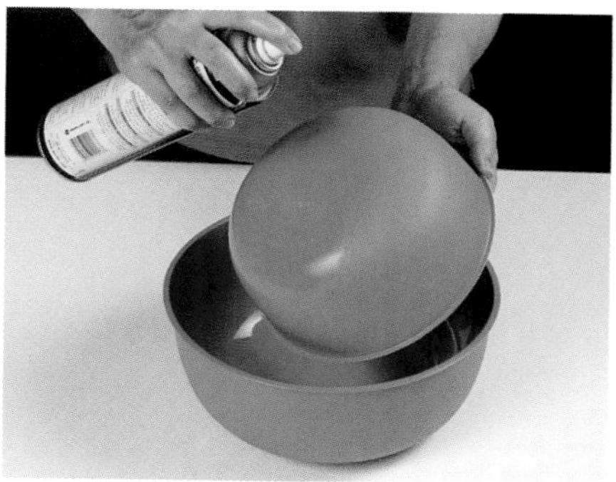

PHOTO 7

PHOTO 8

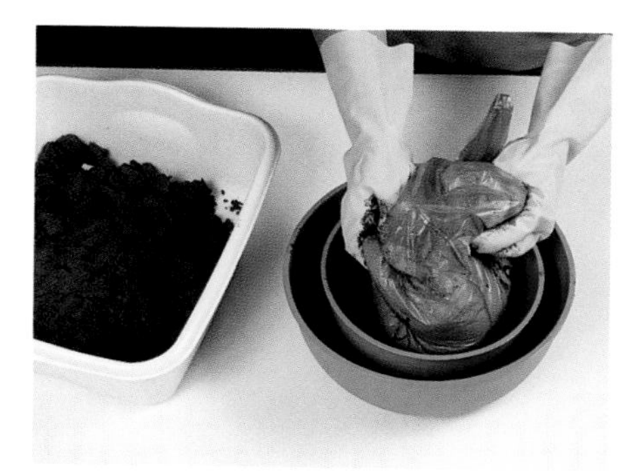

PHOTO 9

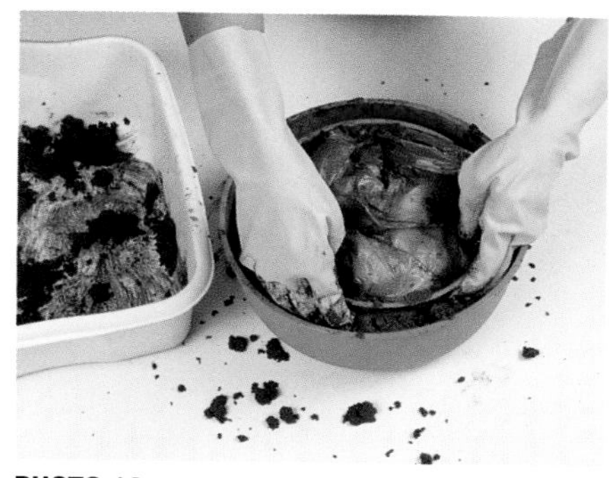

PHOTO 10

PHOTO 11

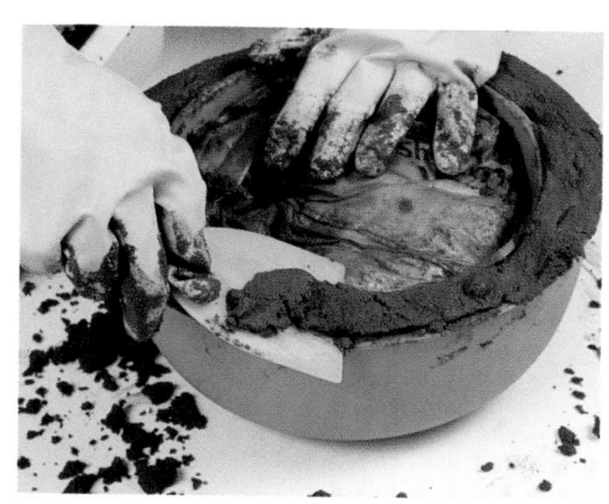

PHOTO 12

PHOTO 13

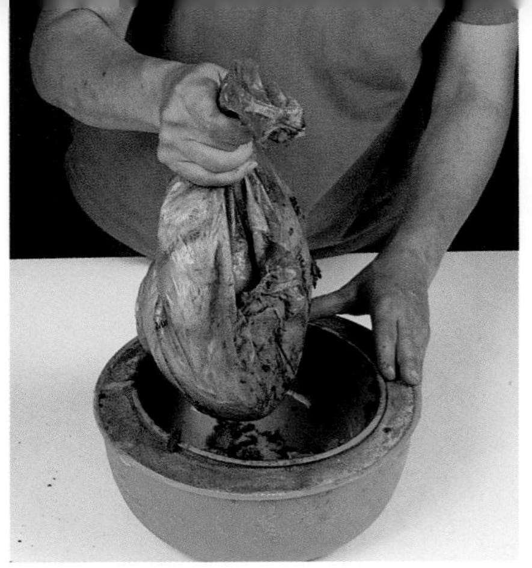

PHOTO 14

PHOTO 15

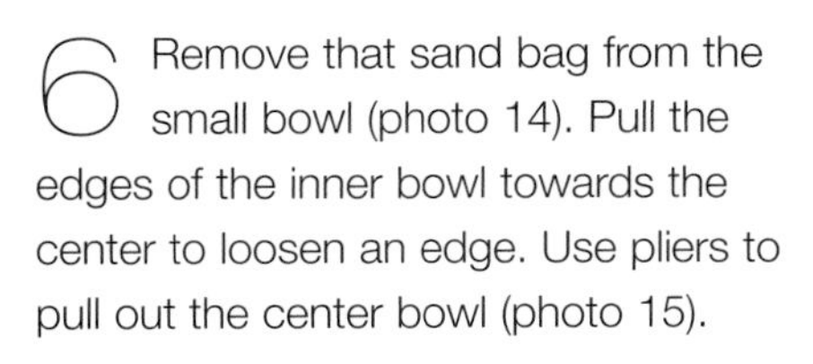

6 Remove that sand bag from the small bowl (photo 14). Pull the edges of the inner bowl towards the center to loosen an edge. Use pliers to pull out the center bowl (photo 15).

7 Loosen the outer edge of the large bowl (photo 16). If needed, use a block of wood to hit the bottom of the bowl (photo 17) a few times to coax the concrete out (photo 18).

NOTE Kem Alexander may have a better idea. She splits her molds first and then puts them back together with tape. To demold, she starts by undoing the tape and removing the two sections.

8 Use a wet sponge and a file to even out the surface and edges (photos 19 and 20).

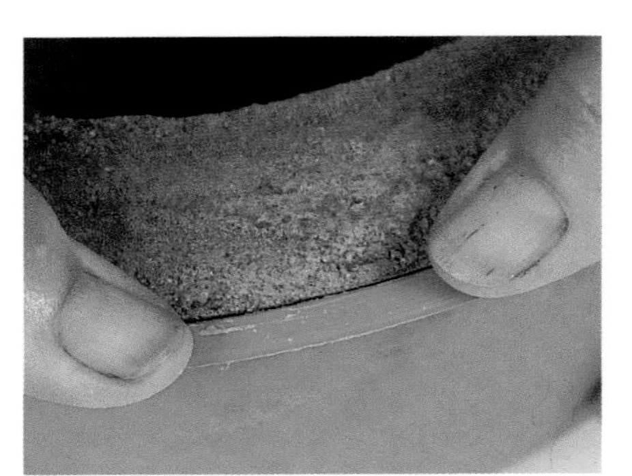

PHOTO 16

PHOTO 17

PHOTO 18

PHOTO 19

PHOTO 20

Kathy Hopwood, *Twins,* 2003. 6 x 6 x ½ inches (15.2 x 15.2 x 1.3 cm). Cement; cast, painted. Photo by artist

Polystyrene Foam Waste Mold

A waste mold is one that's constructed for a single use. In the process of removing the piece from the mold, the mold is destroyed. There are several variations of the polystyrene foam waste mold; many can be made with insulation foam, also known as bead board, that you can find at your home improvement center. Polystyrene foam is so easy to find and inexpensive that it's an ideal material for experimentation.

Materials and Tools

- Concrete Mix: Premixed Commercial-Grade Mason Mix, Mix 1, Mix 2, or Mix 3
- Polystyrene foam insulation
- Tool to cut polystyrene foam
- 3-inch (7.6 cm) nails
- Long screws with large washers (plastic milk tops can be substituted for large washers)
- Permanent marker
- Drywall compound (optional)
- Aluminum roof flashing (optional)
- Spray lubricant (optional)
- Duct tape
- 4 ml plastic
- Nylon scrub brush

Instructions

1 Design the shape you want to make your cast form. Cut out a paper pattern.

2 Cut a section of polystyrene foam insulation large enough so it's at least 1 inch (2.5 cm) larger than the pattern all the way around. Cut as many pieces of the insulation as it takes to form the desired thickness of your design.

3 Pin the sections of polystyrene together with 3-inch (7.6 cm) nails. Position the pattern on the center of the polystyrene foam, and trace it onto the surface with the marker.

4 Cut out the shape through all thicknesses of the polystyrene. Enter and exit through the same cut so the perimeter of the section stays intact. If the cutout of your polystyrene foam is rougher than you want, you can apply drywall compound to the surface to smooth it. Let the compound dry, and sand lightly if needed before casting. No mold release agent is needed for this mold. As a variation, you can also cut strips of aluminum roof flashing and tape them to fit the contour of the opening. When the mold is filled with concrete, the metal will hold tight against the sides. Spray the metal with alight coating of spray lubricant.

5 Secure the layers and edges of the mold with duct tape before casting. Lay the mold on a flat, plastic-covered surface to cast. If you're using this method for a larger mold, you might want to use a few long screws with large flat washers to secure the polystyrene to the surface to prevent the mold from moving as you fill it (see the Garden Bench, figure 1, on page 139).

6 Refer to Casting a Simple Mold or Form, pages 38 through 39, and follow steps 3 through 8 substituting larger reinforcing material for the hardware cloth if making a larger form.

7 Screed or trowel the top surfaces flat. Cover with plastic and allow to sit undisturbed for 24 hours.

8 To remove the mold, take out any nails or screws, remove the duct tape, break the mold away, and discard the polystyrene foam. Take the casting outside and spray it with water. Scrub the surface with the nylon brush to remove any debris. Cover with plastic and cure for one week.

Adapting a Polystyrene Foam Waste Mold for a Totem Construction

Here's a simple variation of the waste-mold technique that will let you stack pieces on a pole to create wonderful totems for the garden.

Additional Materials and Tools

Concrete Mix: Premixed Commercial-Grade Mason Mix, Mix 1, Mix 2, or Mix 3

PVC pipe

Scissors or tin snips (optional)

Saw

File

Instructions

1 Follow steps 1 through 4 as explained on page 43. Position the cutout in the direction you want it to sit on the totem, and cut a piece of PVC pipe 2 inches (5.1 cm) longer than the polystyrene form cutout.

2 For this process, your polystyrene mold needs to be in two halves. Either cut the foam or unpin sections for this step (photo 21).

3 Refer to the instructions for Adapting a Polystyrene Foam Core for a Totem Construction on page 59. Proceed, making this significant modification: instead of inserting the pipe into the core, you'll be inserting it into the waste mold. The mold will hold it in position for casting so it can be incorporated into the final shape (photo 22). If you have lined your mold with the roof flashing, cut holes at the appropriate locations with the tin snips so the pipe can be positioned.

4 Prepare your hardware cloth by cutting two shapes ½ inch (1.3 cm) smaller than the inside of your mold.

5 Mix the concrete. Then position the waste mold on a plastic-covered work board. Fill the mold with 1 inch (2.5 cm) of concrete. Cut one of the hardware cloth pieces in half and position onto the cast concrete, under the pipe.

6 Continue to fill the mold over the pipe, within 1 inch (2.5 cm) of the top of the mold. Lay the other hardware cloth piece on top of the concrete.

7 Follow steps 6 and 8 for the Polystyrene Waste Mold instructions, page 43.

8 Use the saw to trim the pipe even with the casting. Saw the pipe as close as you can without damaging the piece. File the pipe flush.

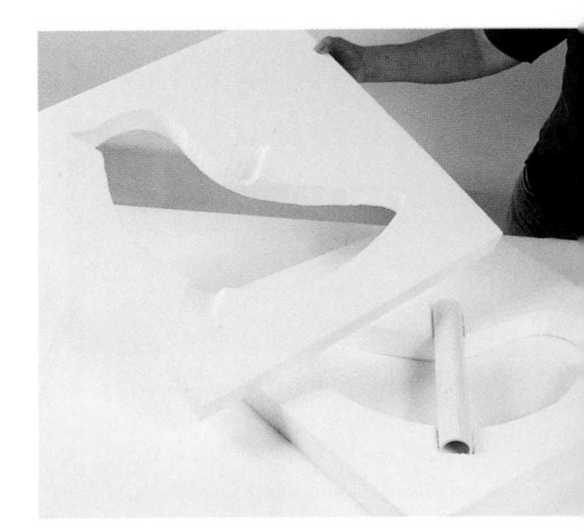

PHOTO 21

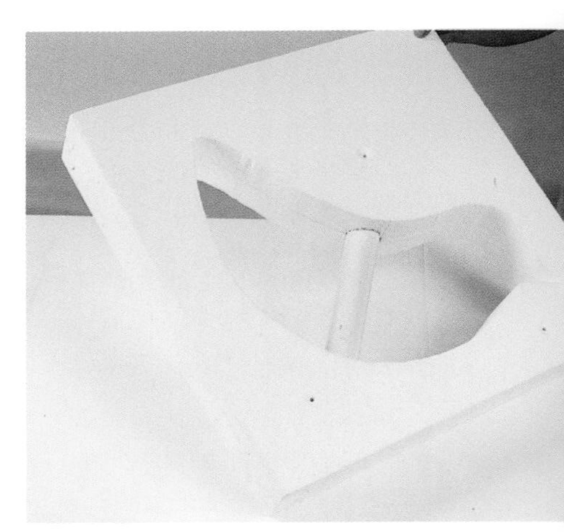

PHOTO 22

Another Polystyrene Waste Mold

Bead board functions well as a frame in the previous examples, but there's also an exciting way to use pink (or blue) expanded foam insulation. This example is just a wall hanging, but don't let your imagination stop here. Artists Marvin and Lilli Ann Rosenberg have used this technique on large forms that were built to cast walls for a skateboard park.

Materials and Tools

- Concrete Mix: Premixed Mortar Mix, Commercial-Grade Mason Mix, or Mix 3
- Insulation foam
- Cutting tool (utility knife, jig saw, band saw)
- Paper for pattern
- Permanent marker
- Sandpaper (optional)
- Soldering iron
- Wood for frame
- Wood fasteners
- Foam adhesive
- Foam tape (optional)
- Mold release (soap works well here)
- Reinforcing material
- Aviator shears
- Container for mixing concrete
- Wire for hanging loops
- Trowel
- Hammer

Instructions

1 Draw your design to make a pattern. Trace the pattern on the foam with a permanent marker (photo 23).

2 Cut out the design using your selected tool (photo 24). Sandpaper can be used to round edges if desired (photo 25).

3 Use the soldering iron to add texture to your forms (photo 26). Experiment with the various tips on scraps to see what kind of effects you can achieve. Try to avoid making lines that are too deep or narrow. Your concrete may not release as well from these areas when you remove the foam.

PHOTO 23

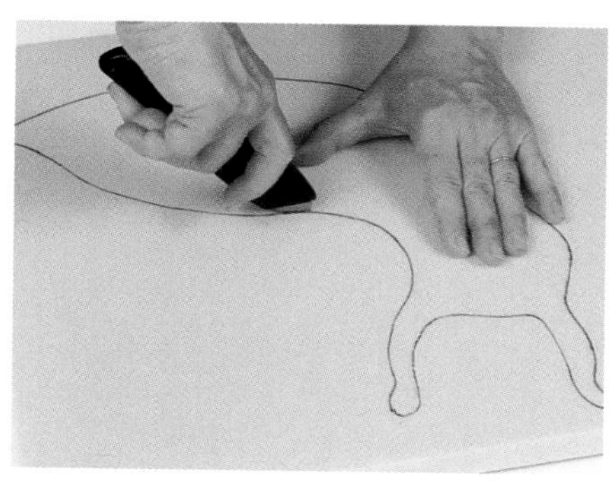

PHOTO 24

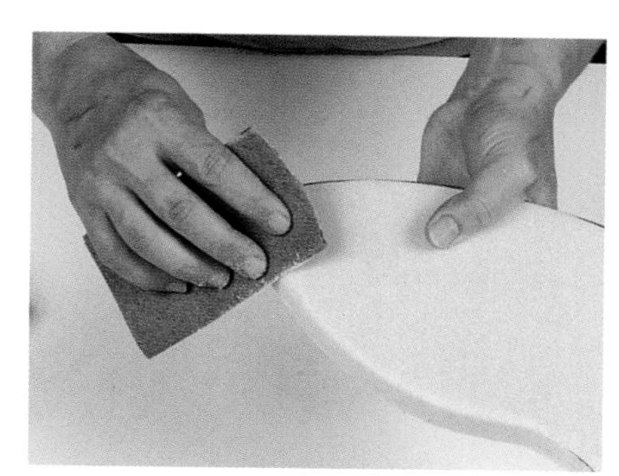

PHOTO 25

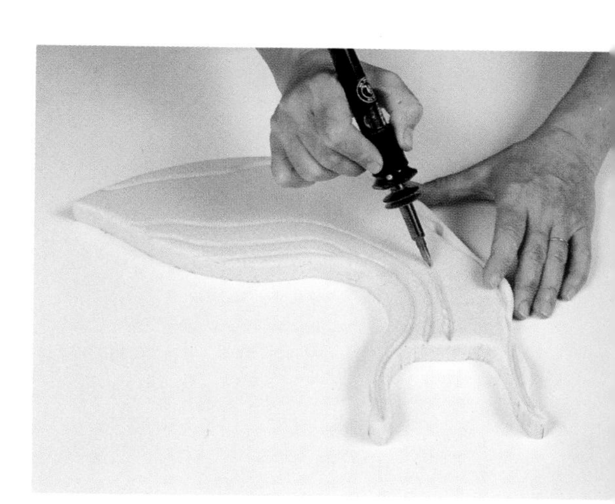

PHOTO 26

4 Assemble your design in the frame, starting with the placement of a piece of insulation board. This will act as the foundation for the mold (photos 27 to 29). Apply release agents to the wooden frame and to the foam (photo 30).

5 Mix your concrete. Begin to apply it in small handfuls onto the textured areas (photo 31). Continue to fill the frame until it is half full and then add in your reinforcing material.

PHOTO 27

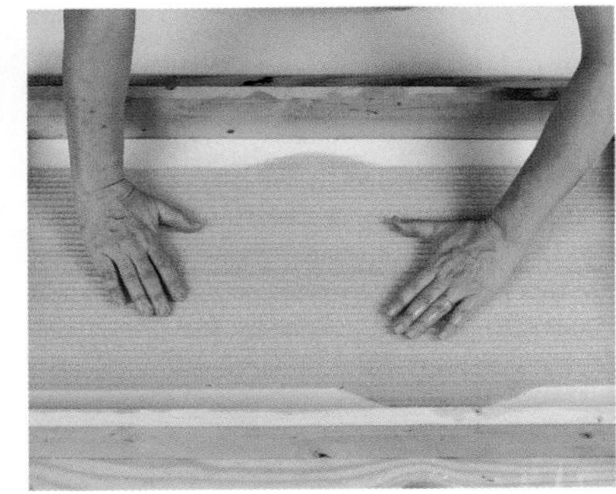

PHOTO 28

PHOTO 29

PHOTO 30

Ricky Boscarino, *Monolith at the Ganesh Temple,* 2001. 10 feet (3 m) in height. Concrete, glass; mold cast.
Photo by artist

PHOTO 31

PHOTO 32

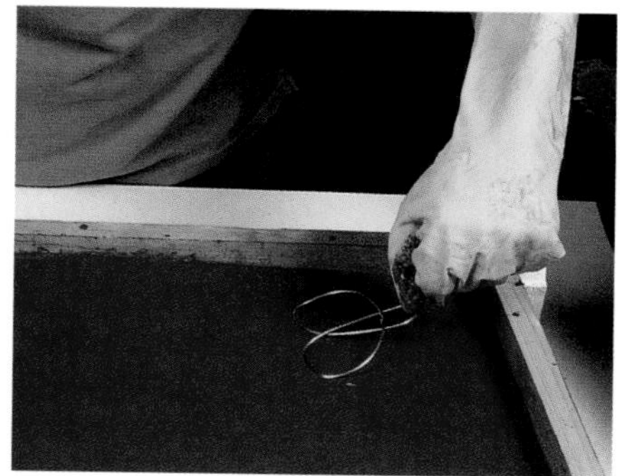

PHOTO 33

PHOTO 34

PHOTO 35

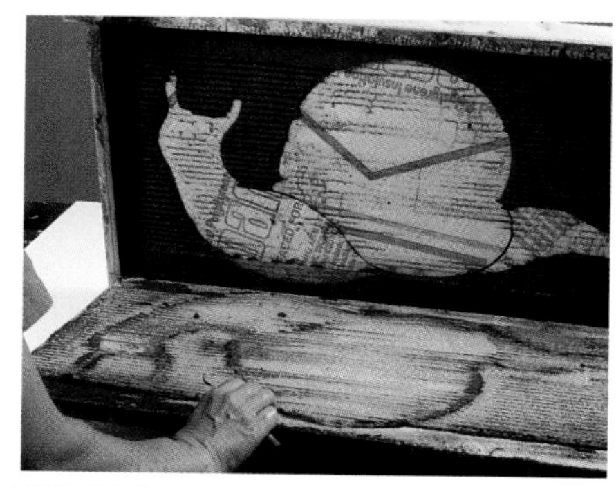

PHOTO 36

PHOTO 37

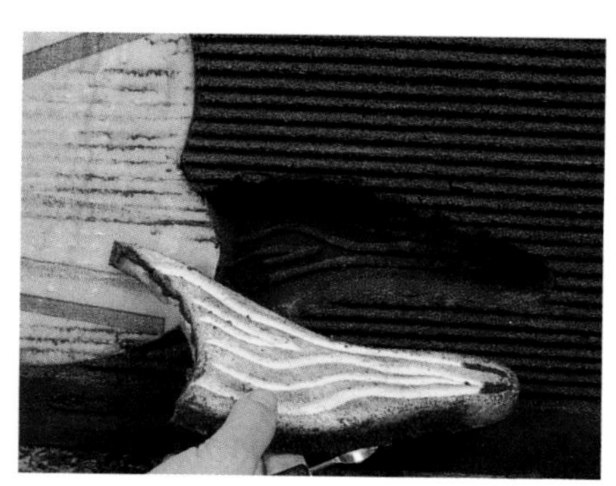

PHOTO 38

6 Trowel the top surface smooth (photo 32). Insert the hanging loops (photo 33) and make sure they are well set into the concrete (photo 34).

7 Hammer the outside of the frame to remove bubbles (photo 35). Cover the piece with plastic and let sit undisturbed over night.

8 The next day, carefully turn your casting so you can work from the bottom (photo 36) and begin to remove the foam. If any concrete has migrated over the edges of the foam sections, remove it with a sharp tool (photo 37). If you try to remove the foam with concrete on the edges, you run the risk of chipping the casting.

9 Finish removing your foam (photo 38). Lay the casting face up on the boards so the hanging loops are not against the work surface. Unscrew your wood frame, and file any sharp edges. Cover the piece with plastic for at least 5 days to cure.

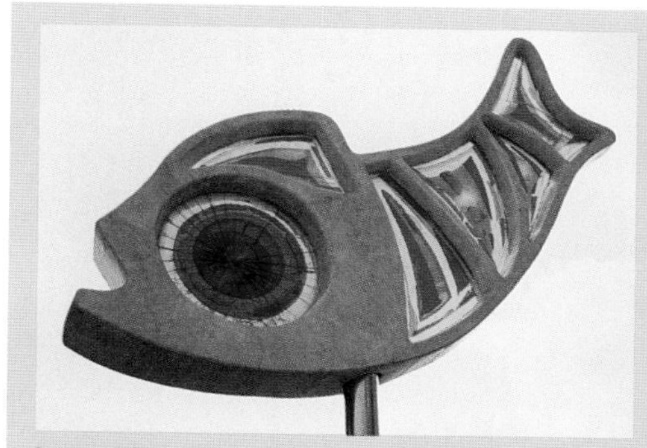

Lynn Olson, *Flying Fish,* 2003. 36 x 24 x 3 inches (91.4 x 61 x 7.6 cm) on a 6-foot (2.2 m) stainless steel pole. White cement, stainless steel rods, carbon fibers, stained glass, latex polymer; direct cement construction, surface filed, sealed with methyl methacrylate. Photo by artist

Sand Casting

Sand casting is one of the easiest and most economical ways to get started with casting a variety of shapes or forms. You can create decorative wall hangings by simply casting into indented shapes in sand. I primarily use the sand to form mounds, casting onto them. Variations on the sand-cast mask technique below can produce everything from birdbaths and fountains to planters. Larger versions of this technique have been utilized to create architectural structures. Instead of sand, large dirt mounds are formed, a reinforcing structure is constructed, and the concrete is applied. When hardened, the dirt is dug out to reveal an interior space.

Materials and Tools

Concrete Mix: Premixed Sand/Topping Mix, Commercial-Grade Masons Mix, Mix 3, Mix 5, or Mix 12

4 ml plastic

Approximately 2 gallons (7.6 L) sand

Plastic tub

Water

Dry-cleaning plastic bags or thin plastic drop cloths

Permanent marker (optional)

Aviator shears

Reinforcing material (hardware cloth or AR fiberglass cloth)

Wire for armatures

Angle cutters

Needle-nose pliers

Container to mix concrete

Putty knife

Brush

Large plastic trash bag

Coarse rasp or file

PHOTO 39

PHOTO 40

Instructions

1 Cover your work area with 4 ml plastic. Place approximately 1½ gallons (5.7 L) of sand in a tub.

2 Mix enough water into the sand so it packs easily (photo 39). Remember the last time you made a sand castle at the beach? That's the consistency you're looking for.

3 Place the sand toward the middle of your work area so the finished form will be at least 4 inches (10 cm) from any edge of the table or work board. Form a shape that will be the inside shape of your mask (photo 40). The height of your mound will be the depth of your mask. (If you're making a bowl, the shape of the sand will be the inside of the bowl since you'll be casting in reverse.) Your mask can be any shape; this example is quite different from the project mask on page 135, but the process is the same. Experiment with the contour of your

PHOTO 41

mound by adding and subtracting the amount of sand (photo 41). Be sure to pack your sand shape firmly so it won't collapse when you apply the concrete.

4 Clear off excess sand and dry the surrounding plastic. At this point, you might want to use the permanent marker for drawing a line on the plastic to define the shape you want for the outside edge of the mask; it can be a completely different shape than your sand mound. This line can also serve as a pattern for cutting your reinforcing materials.

5 Cover the sand form with a single layer of thin plastic that extends several inches past your mound (photo 42).

6 Cut the reinforcing material with the aviator shears to the shape and size of the finished mask. Your reinforcing materials will be larger than the sand form, but not larger than the finished mask. Set aside.

7 Since this mask has ears and was designed to hang on a wall, make small armatures. They will add support to the ears, integrate them into the structure, and provide loops for hanging. Use stainless steel or galvanized wire to make two armatures, as shown in photo 43.

8 Position the ears and poke the loops through the plastic, into the sand (photo 44). Remember, the sand creates the open space behind the mask. You want the loops to be sticking out of the concrete surface. Mix the concrete. Use small pinches of concrete and work it into the ear armatures (photo 45).

9 Take a handful of concrete and form it into a ¾-inch-thick (1.9 cm) "hamburger patty." Lay the patty in the middle of your form.

PHOTO 42

PHOTO 43

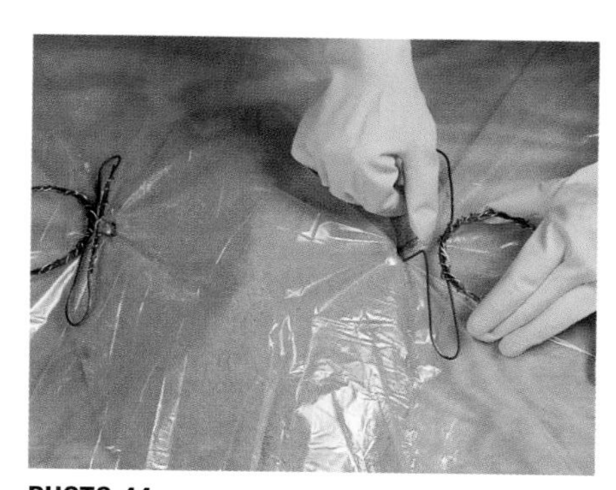

PHOTO 44

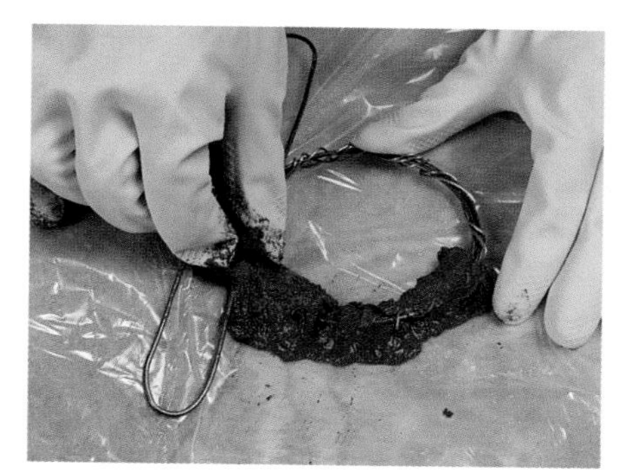

PHOTO 45

10 Make another patty, and place it so it overlaps the first by at least ½ inch (1.3 cm). Pat lightly until the seam line disappears (photo 46).

11 Continue applying the concrete in this manner until the form is covered to an even thickness of approximately ½ inch (1.3 cm), as shown in photo 47. Notice that the look of the mask is different from the original sand-cast form. Use the putty knife to trim the concrete to refine the finished shape.

12 Place the reinforcing material over the concrete and press in lightly to embed (photo 48). You may need to add a small amount of concrete over the surface to secure the material to the mask (photo 49). Don't worry if some of the reinforcing material shows through.

13 Use a sponge to help shape the ears (photo 50). Pull the edges of the thin plastic toward the center of the casting (photo 51). This will help you coax the concrete back—these things have a way of growing as you work—and compress the material at the edge, making it stronger. Leave the plastic in the drawn-up position.

14 Cover the piece with an unopened trash bag, and leave undisturbed overnight (see the photo at top left on page 25). The next morning, remove the trash bag and pull back the thin plastic.

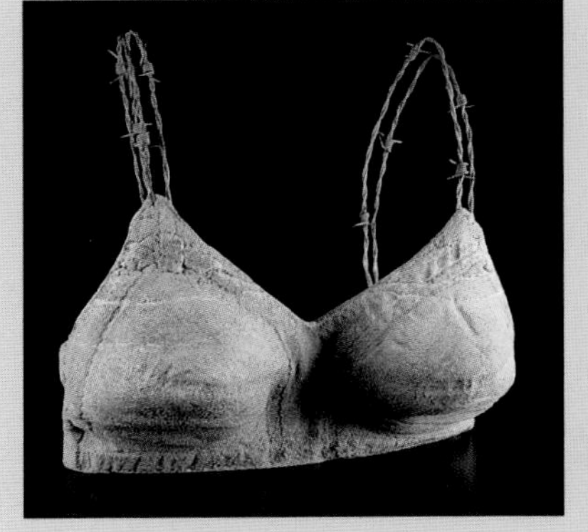

Kem Alexander, *The Bra,* 2004. 12 x 11 x 7 inches (30.5 x 27.9 x 17.8 cm). Concrete, old bra, galvanized barbed wire; hand packed, sculpted, cleaned.
Photo by Art Image Studio

PHOTO 46

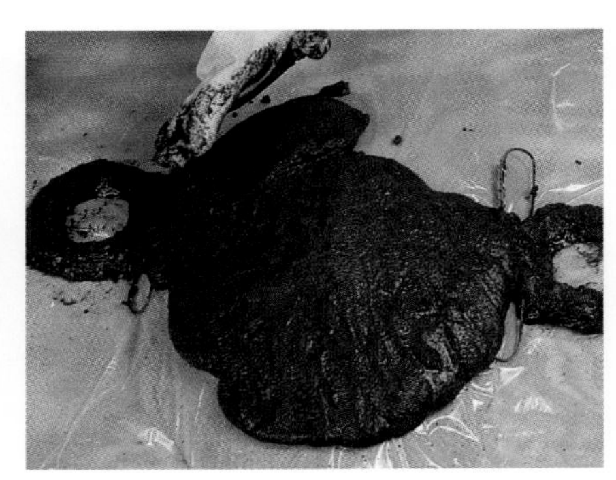

PHOTO 47

PHOTO 48

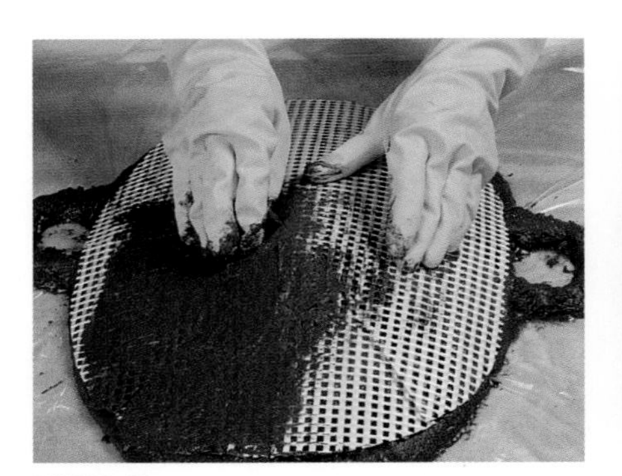

PHOTO 49

PHOTO 50

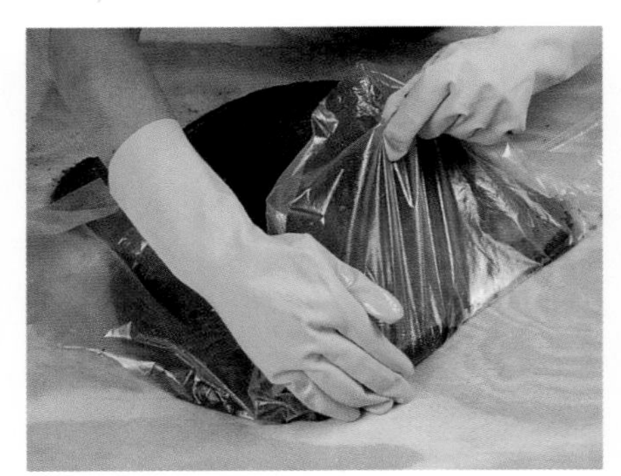

PHOTO 51

15 Carefully pick up the edge of your mask, and lift it off of the sand. It's easier to pick up the edges if you put one hand under the plastic that's covering the table. Lift the mask from underneath, and grab the edges with the other hand (photo 52).

16 Peel the thin plastic from underneath your mask (photo 53). You'll see how the loop is sticking out.

17 Carefully file any thin or loose sections, and define the shape as needed (photo 54). Spray with water, cover in plastic, and let the mask cure for at least three more days.

NOTE The instructions for this cast mask just create a base form. To finish the mask, you'll want to add more concrete to the surface to add more strength and to refine the image. Refer to Chapter 3 to explore options for surface treatments.

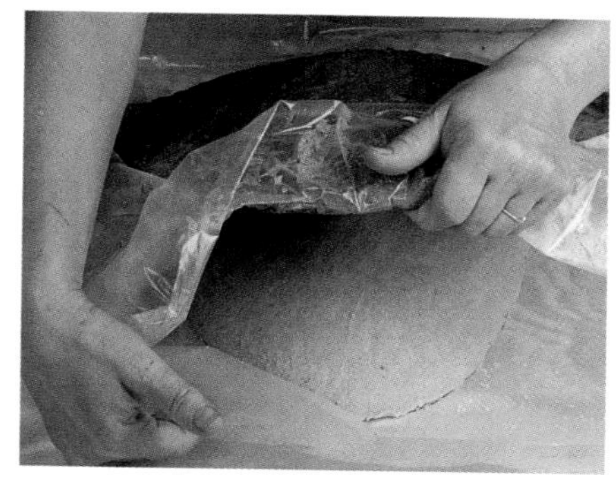

PHOTO 52

PHOTO 53

Johan Hagaman, *Covered in Birds,* 2004. 26 x 5 x 5 inches (66 x 12.7 x 12.7 cm). Hand formed concrete over steel, found objects, handmade objects, paint; acid stained. Photo by Richard H. Connors

PHOTO 54

Bowl Connector

For some of your constructions, such as the Birdbath on page 144, you'll want to add a stem to the bottom of a cast bowl so you can connect it to a base. A plastic drinking cup provides a simple and effective means of creating a connecting system.

Additional Materials and Tools

16-ounce (480 ml) plastic cup

Utility knife

Needle-nose pliers

Instructions

1 Cast a bowl form adapting the Sand-Casting Instructions, pages 48 through 51. Proceed with the following steps after you position the reinforcing material and cover it with concrete, as in step 12 on page 50.

2 Determine where the center of balance of the bowl is located and make a mark in the wet concrete.

3 Fill the cup with concrete, tamping it as you go, to avoid air bubbles. Fill the cup completely, and then add a little more so it resembles a glass of soda when the foam crests slightly above the rim of the glass.

4 Turn the cup over and place it at the marked point. Give the cup a gentle twist. Eye it from several angles so you're sure it's straight, and then leave the cup in place (photo 55). Pull the edges of the thin plastic toward the center of the casting. Cover with a piece of plastic and leave over night.

5 The next morning remove the plastic. With the utility knife, carefully score the plastic cup. Use the needle-nose pliers to grab and remove the cup (photo 56). Continue with steps 15 through 17 on page 51 to complete your form.

Elder Jones, *Fountain,* 2004. 10 feet (3 m) in height. Concrete; wet carved.
Photo by artist. Collection of Linda Mattson, Nashville

PHOTO 55

PHOTO 56

Flexible Mold

If you want to make duplicate castings of an object that's already made in a different material, you may want to make a flexible mold. This process can be a little tricky and requires different specialized materials than most of the projects in this book. However, improved technology in recent years has definitely simplified the mold-making process. Previously, almost all mold-making materials had to be carefully weighed, which required an involved set up. Now it's easy to find a good working system that only needs to be measured by volume in equal parts.

When you're purchasing your mold-making materials, there are two initial considerations—the model you're making the mold from, and the material you will be casting in the mold. Mold-making materials are available in brush-on, sprayed, or poured systems. Here again, your model will help determine the application.

A polyurethane or urethane liquid rubber is recommended for making molds for casting concrete. One that exudes an oil, or that's referred to as *wet*, aids in the demolding of concrete. You'll find that different rubbers have varying degrees of flexibility. The amount of detail in your model will have some bearing on how flexible you'll want your mold.

Materials and Tools

- Plasticine/modeling clay
- Work board (for bas-relief base)
- Assorted clay tools (optional)
- Acrylic work board (for casting)
- Sealing agent
- Release agent
- Containment frame (wood, plastic, metal or cardboard for smaller pieces)
- 2-part urethane mold system
- Disposable mixing containers
- Stirring sticks
- Latex gloves

Instructions

1 Select or make a model for casting (photo 57). I wanted to make a panel, which will work well with the poured method. A three-dimensional piece or even a bas-relief with more undercuts might require brush or spray-on applications.

2 Find a level work area. Working on a sheet of acrylic will help you remove your mold easier. Using some of the modeling clay, seal the edges of your containment wall to the acrylic. Seal the edges of your model to the work surface as well.

3 If your model is porous (plaster, stone, wood, concrete) or made of clay, it will need to be sealed. I was told that water-based clay, when used as a model, sometimes reacts to the urethane rubber causing surface bubbles. Since then, I've been using plasticine. There are different sealers readily available such as shellac, spray acrylic, paste wax, or petroleum jelly thinned with mineral spirits. There are also specific commercial products that are made and recommended by the manufacturers of mold-making systems.

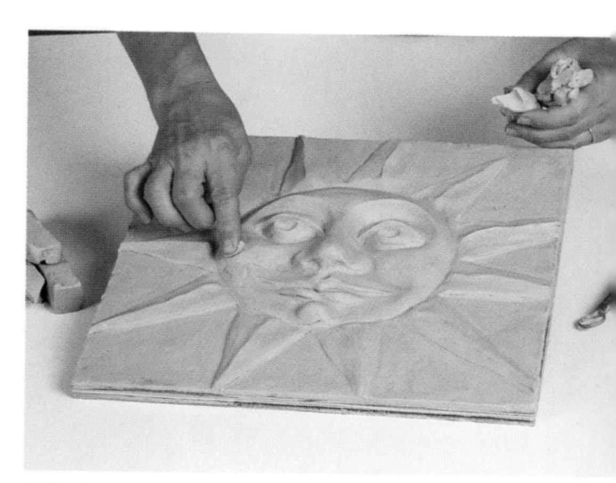

PHOTO 57

PHOTO 58

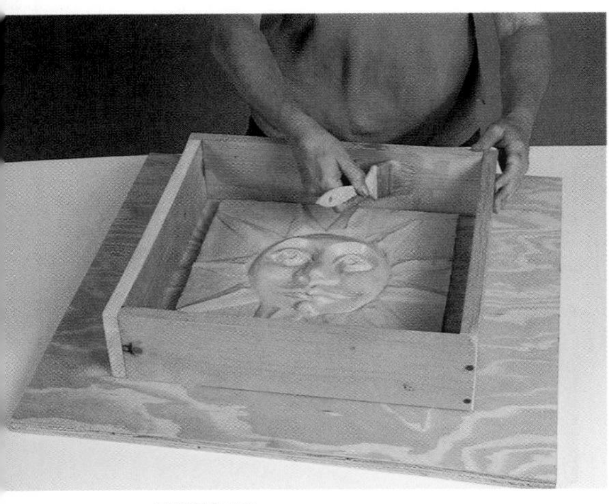

PHOTO 59

PHOTO 60

Apply one coat of the sealer in one direction (photo 58), let dry, and then apply another coat in an opposite direct to ensure good coverage.

4 The next step is to apply a recommended release agent to the model.

NOTE These are not the same release agents we have been using in our concrete molds. If you do not use the correct release agent on your model when making a rubber mold, it may never come off! Apply two light coats in the same manner as the sealer. Remember to use sealers and release agents on your containment frame (photo 59), and on every surface the rubber will come in contact with. Be careful though; too heavy of an application may result in surface bubbles.

5 Thoroughly mix the urethane rubber (photo 60). Mix for at least three minutes. The most common reason for a mold not to cure properly is improper mixing. Avoid introducing air bubbles into the mix by mixing slowly and scraping the sides and bottom of the container while stirring.

6 In this example, I'm making a poured mold. Do not pour the rubber over the model. Find the lowest point of the mold frame and pour the rubber in that one spot at a continuous, slow rate. Allow the mold-making material to flow naturally to fill in areas without trapping air. Fill the area until the highest point of your model is well covered by the rubber.

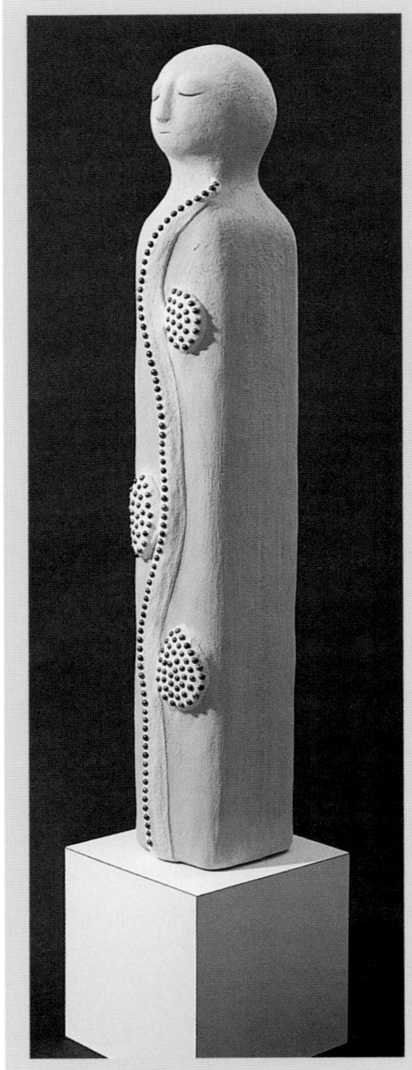

Kathy Hopwood, *Stillness,* 2003. 47 x 8 x 8 inches (119.4 x 20.3 x 20.3 cm). Polymer fortified cement with alkali resistant fiberglass mesh, polystyrene foam core form, marbles, grout. Photo by Andrew Ross

7 Allow the mold to cure overnight or for about 16 hours.

8 Demold your model. If everything has gone well, it should peel off your model and show all the detail of the original.

Armature material, left to right: polystyrene, rebar, angled galvanized slotted metal, flat galvanized slotted metal, shelving bracket

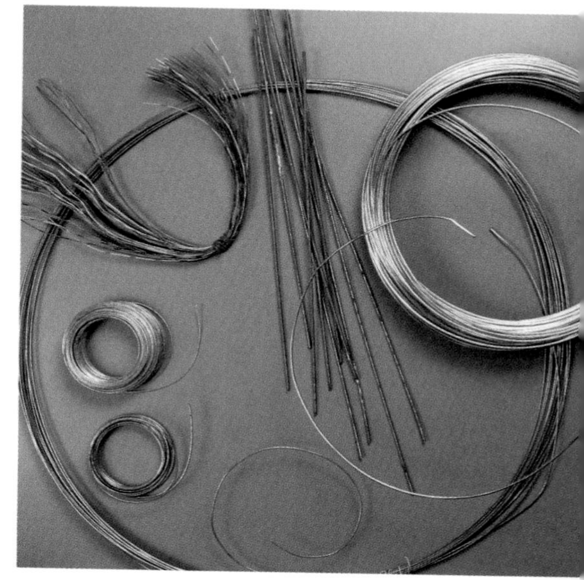

Wire for armatures, clockwise from top: insulation hanging rods, 16-gauge galvanized wire (clothesline wire), 20-gauge galvanized wire, 22-gauge galvanized wire, plastic-coated wire (telephone wire). Background: large ring of stainless steel wire

Armatures

An armature is the skeleton of a sculpture—the inner support. Whether you want to create a standing figure or animal, or an abstract form that swoops skyward, an armature is the structural frame that will allow you to apply heavy or soft materials to make those forms possible.

As you start a sculptural form, you need to think of it from the inside out. I'm not suggesting that you work out every detail of the piece before you start—where's the excitement in that! But it does help if you start with an understanding of the basic gesture (movement) and the scale (size) of the piece.

The armature needs to be contained within the piece unless you have specifically intended for it to be part of the finished work (see Johan Hagaman's work on page 51). The finished armature must also be smaller in volume than you envision your completed work. A bulky armature will only get bulkier as you apply your sculpting medium to it.

Keep in mind that your finished concrete piece doesn't have to be solid to be sturdy. The materials you choose for your armature can lighten the completed piece. A cardboard tube or taped form helps to create air space, while a large piece of polystyrene foam will provide a significantly lighter core. Even using newspaper to stuff a wire form before applying the concrete will keep the concrete from filling the entire void.

The armature will work more effectively if you choose the right material for its construction. You'll want to select heavier or stronger materials for constructing armatures for larger projects than you'll need for small or more compact pieces. The following examples will provide you with information for some basic armature constructions. Use them as reference for the projects as well as to guide you as you experiment with your own sculptural explorations.

Polystyrene Foam Armatures

Polystyrene foam is a versatile material that can be used to create both geometric and organic forms. I use the type of foam made up of compressed beads rather than the expanded foam. It's available as insulation at some

Tools for carving polystyrene, clockwise from top: assorted sand paper, hot knife, assorted saws (folding saw, keyhole saw, pruning saw, drywall saw), caulk gun with construction adhesive, spray adhesive, assorted wire brushes

home improvement centers and is referred to as bead board. The compressed-bead polystyrene is rated in pounds: 1, 2, and 3 lbs (.45, .90, and 1.8 kg). The numbers have to do with the pressure used to compress the beads—the higher the number the harder the foam.

I like to carve the 1-pound foam. It's dense, but you can easily use the tools suggested for carving. To locate polystyrene foam in your area look in the phone directory under "Packaging," or try to find a building supply company that sells polystyrene foam for boat docks or even archery targets. I've had more success carving out the shapes myself, but on occasion I'll use small premade forms available at hobby shops

When you use polystyrene foam as an armature it's also useful to think of it as a core. In some cases you'll still need to reinforce your polystyrene shape or you may need to build blocks of polystyrene around a metal armature before carving. This is especially true if you're working on larger pieces that might invite climbing or sitting. As a core armature, polystyrene is ideal to use when applying a polymer-fortified concrete and alkali-resistant fiberglass system. (See pages 82 through 83 for a complete description of this process.) The layering of these materials conforms to almost any shape. In fact, applications of similar materials can frequently be seen in modern-day architectural building systems. As with other armature constructions, you should carve your polystyrene foam smaller than you envision the finished piece so the finished piece isn't bulkier than you want it to be.

Johan Hagaman, *Kissing Bees,* 2003. 25 x 5 x 5 inches (63.5 x 12.7 x 12.7 cm). Hand formed concrete over steel, found objects, handmade objects, paint; acid stained. Photo by Richard H. Connors

Carving a Polystyrene Foam Armature

I learned how to carve foam by looking at woodcarving books. Regardless of the material, the basics are the same even if the tools are different. Refer to Carving Basics on page 76.

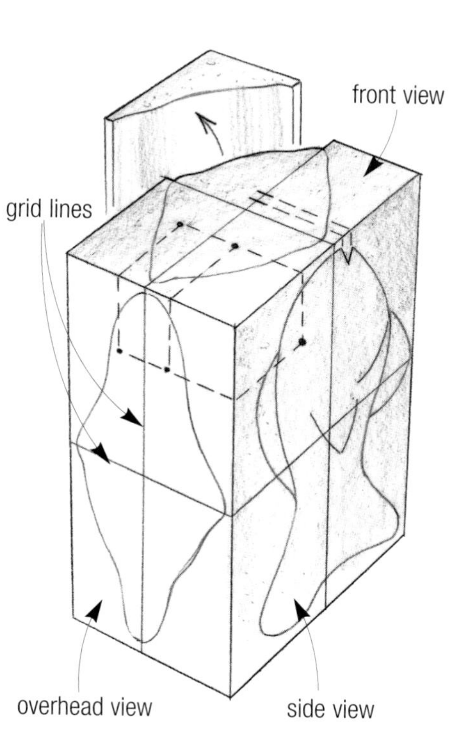

FIGURE 2. Organizing the way you draw the different views of your sculpture will help you transfer the location of the details.

Instructions

1 Design your piece to scale either by making a sketch or model. Draw the scale grid onto your polystyrene foam block, as shown in figure 2.

2 Use a pattern or draw freehand onto the scaled block to make the profile outline of the design for the side view and front view, and for the overhead view if your design has one (photo 61).

3 Use your selected tool to cut out the side view (photo 62). Try to make your cuts so the scraps remain as intact as possible. Here I'm using a homemade hot-wire table.

NOTE If you're using a hot wire, be sure to work in a well-ventilated space. You want to avoid breathing the toxic fumes from the melting foam.

4 After you've cut out the side view, pin the scraps back onto the block using large common nails so you can see your drawing on all sides again (photo 63). Take care not to place the nails in the line of your next cut.

Materials and Tools

- Polystyrene foam block
- Permanent marker
- Cutting tools: hand saws, hot knife, hot wire, and band saw
- Large common nails
- Assorted wire brushes

PHOTO 61

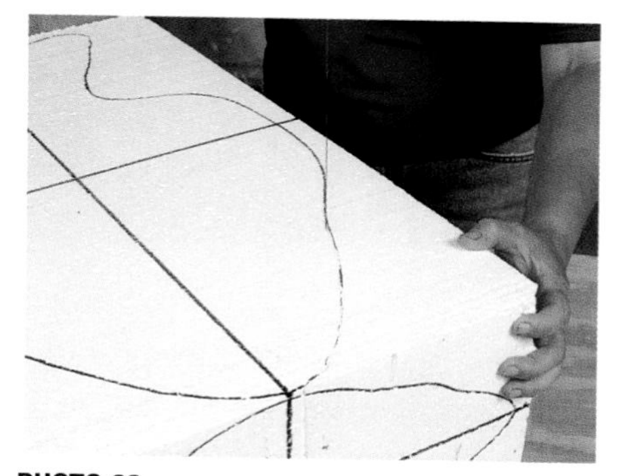

PHOTO 62

PHOTO 63

PHOTO 64

PHOTO 65

PHOTO 66

PHOTO 67

5 Turn the block to follow the lines of the front view as you cut the block again (photo 64 on page 57).

6 If you're using a third view, pin the scraps together again to reform the original block and make your third set of cuts. When you've removed the nails and let the scraps fall away, you'll have a blocky likeness of your design (photo 65).

7 Additional cuts can be made on the pieces as needed (photo 66). Round the edges and define the form using the wire brushes (photo 67).

This basic carving technique works even on larger pieces. If you don't have a cutting tool available to make profile cuts, your approach will be different but still successful—it just may take a little longer. Starting with your initial design and scaling as described in steps 1 and 2, use the cutting tool you have available to remove edges, planes, and large chunks. Use your wire brushes more aggressively to remove more material rather than to just refine the surface. A wire wheel attachment on the end of a drill can be a big time saver, but it only works on larger pieces of foam or on smaller pieces that can be secured. Otherwise, you'll just end up chasing your foam because the power of the spinning wheel wants to push your piece away. Wear goggles and a dust mask when you work to avoid getting polystyrene in you eyes or up your nose. I can attest to the discomfort they can cause.

Continually assess your carving progress. As you work, use a permanent marker to draw on your form to re-establish guidelines. If you're working on a symmetrical piece, continually check your measurements; if you're creating a seating unit, sit in it to see how it feels. Take the time to view your piece from all angles.

How do you know you've carved enough? I like to carve on a piece of foam until it's the shape of the design I've envisioned. Then I review how I intend for the piece to be used and the final surface treatment it will receive. If I'm using the polystyrene as a core for a ferro cement construction, I know that I'll be building up a layered surface of metal mesh and concrete that might be 1½ to 2 inches (3.5 to 5 cm) thick. If I'm using the polystyrene as a base for applying a polymer-fortified concrete and alkali-resistant fiberglass system, I'll probably be building up about ¼ to ¾ inch (6 mm to 1.9 cm) of material. If I'm going to mosaic a surface, there's another ¼ of an inch. By totaling the expected heights of the various processes, I get an idea of how much more material I'll need to remove from my carving so it will retain the shape I have in mind without the finished piece becoming too massive.

A polystyrene foam armature can be easily adapted or modified. If you need to add supports to it or are building foam blocks around a metal armature to add structural strength, you can cut the foam and carve tight-fitting cannels into it using small brushes, rasps, and rifflers. Foam adhesives, either commercial sprays or construction caulk adhesive, can be used to construct or reconstruct the foam into a solid form. These adhesives are also useful if you've inadvertently carved off too much or want to change your carving. I always have difficulty carving ears for animals or people in the correct place. I frequently have to use a cutting tool to cut the ears off so I can glue them back on at the correct location. When I do this I first reposition them with large common nails until I'm sure they're in the correct spot, trace around them with the permanent marker, apply the adhesive, and then reattach the ears.

Adapting a Polystyrene Foam Core for a Totem Construction

A totem construction is an easy way to create a larger, more complex project by combining small manageable components. Or, stated another way, you make a series of small pieces that you stack together to make a large piece. One way to approach a totem is to start with a basic theme to help guide you in the design of components or elements (see the photos on pages 18 and 21). You can begin making your totem by using the preceding instructions to carve a series of polystyrene foam shapes. The following instructions will help you prepare the shapes for your totem before you start applying the concrete.

Instructions

1 Take your finished carved-foam core and draw a centerline to guide you as you cut the piece in half. See pages 57 through 58 for instruction on carving polystyrene.

2 With the cut halves facing up, decide how you'd like to position the piece on the pole, taking into consideration the weight of the finished piece and how it will physically balance on the totem.

3 Measure the length of the area where the pipe will be inserted and add 2 inches (5 cm) to the length. Cut the pipe to that length.

4 Measure and mark the outside width of your PVC pipe on the polystyrene foam.

5 Use the small wire brush or riffler rasp to carve out half the circumference of the pipe on one half of the form (photo 68).

6 Hold the two halves together. Use the permanent marker to mark the opening edges of the carved half, transferring the measurements onto the opposite side. Lay the sections down and connect the marks using the straight edge.

7 Carve out the second half as in step 5. When you put the two pieces together with the pipe inserted, the pieces should fit together tightly and the pipe should be snug (photo 69).

8 Glue the pieces together. Apply a generous bead of caulk to both sides of the form and the pipe. Press them together and then immediately pull them apart. Let the adhesive dry slightly until it becomes tacky, which should take about 10 minutes (photo 70). Reposition the sections and then press them together. This contact method makes a stronger initial bond.

9 Secure the edges with a few large nails for added strength.

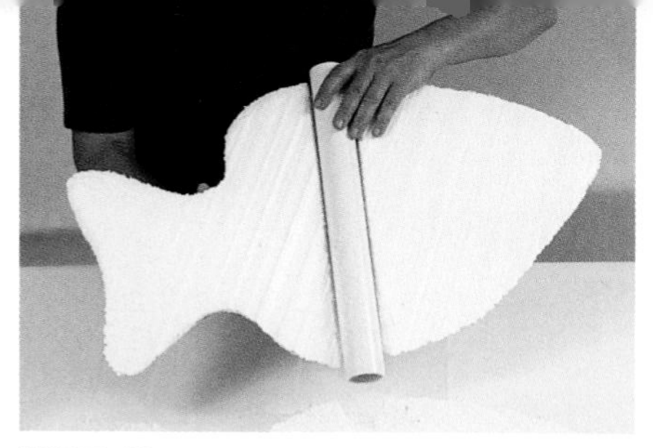

PHOTO 68

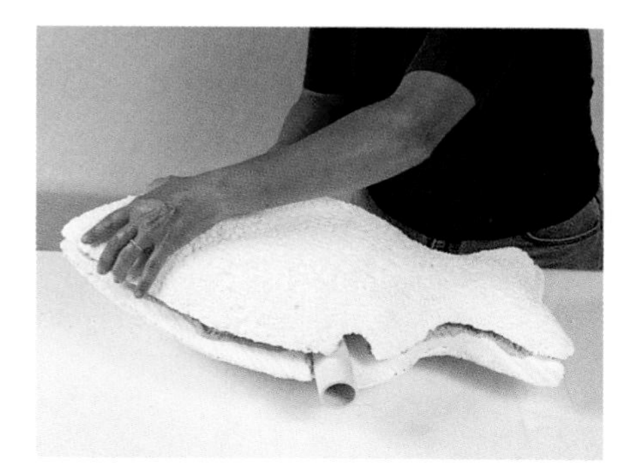

PHOTO 69

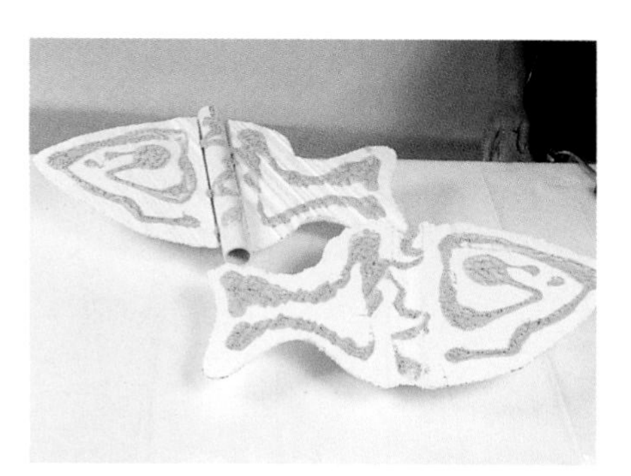

PHOTO 70

Additional Materials and Tools

PVC pipe *

Straightedge

Permanent marker

Small wire brush or riffler rasp

Foam adhesive: used here, construction caulk adhesive

Large construction nails

* When choosing your pipe, take into consideration the inside diameter. Make sure that it will fit over the heavy metal pipe that you will use for sliding the pieces on during installation.

Adapting a Polystyrene Foam Core as a Pole Topper

Sometimes you may want to make a polystyrene foam core construction that sits on top of a pole. However, you can't just stick a pole in polystyrene and expect it to stay there. It will eventually work itself through the foam and possibly come through the top of your form.

Additional Materials and Tools

PVC pipe*

Machine screw and nut

Drill

Drillbit

Straightedge

Permanent marker

Small wire brush or riffler rasp

Foam adhesive: used here, construction caulk adhesive

Large construction nails

* When choosing your pipe, take into consideration the inside diameter. Make sure that it will fit over the heavy metal pipe that you will use for sliding the pieces on during installation.

Instructions

1 You can either start with a finished carving as we did for Adapting a Polystyrene Foam Core for a Totem Construction on page 59, or you can add the PCV pipe before carving.

2 Choose a piece of pipe that is three-quarters as long as your form, and add 2 inches (5 cm) to that measurement. Select a machine screw that is about 1 inch (2.5 cm) longer than the width of the pipe and then a corresponding nut. Select a drill bit slightly larger than the diameter of the screw. Drilll straight through both walls of the pipe, approximately 1 inch (2.5 cm) from the top edge. Insert the screw through the pipe and tighten the nut (photo 71). This screw will stop the pipe from working its way through the top.

NOTE Two other good solutions were recently suggested to me. If you have PVC adhesive, you could buy a corresponding cap and glue it on. Depending on the size of pipe you use for installation, a piece of threaded plumbing pipe with a screw on cap would also work.

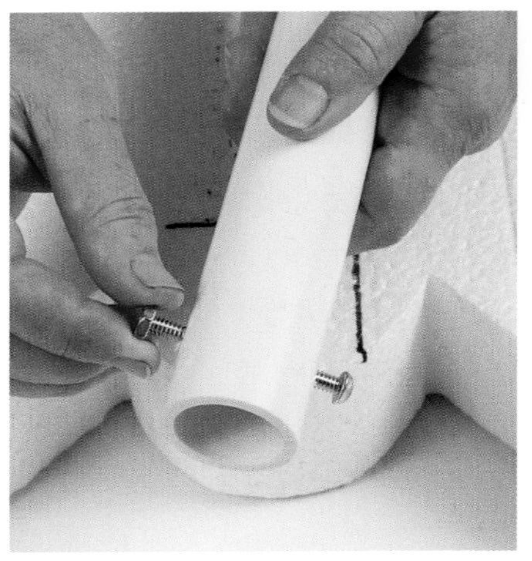

PHOTO 71

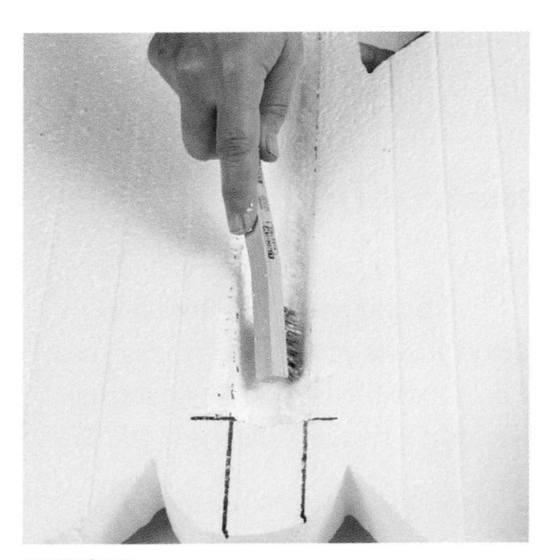

PHOTO 72

3 Orient your pipe in the direction you want the form to sit on the pole, taking into consideration that a mid or balance point works best. Measure the width of your pipe and draw two lines the length of the form.

4 Measure about 2 inches (5 cm) from the top and mark. At that mark, use the wire brush to start carving a space that is equal to half of the pipe's circumference (photo 72). The pipe should fit snug in the carved area (photo 73).

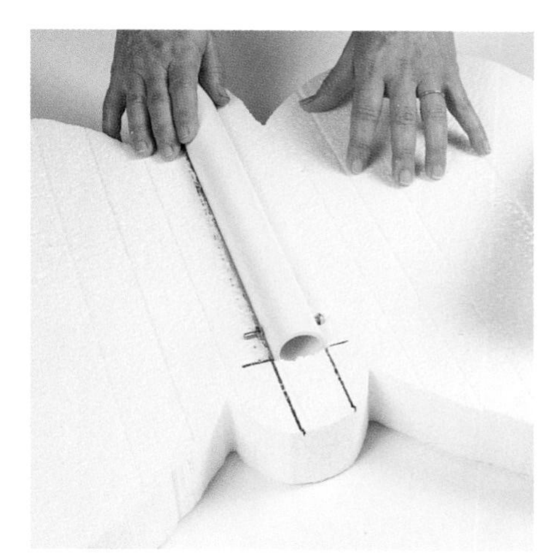

PHOTO 73

PHOTO 74

5 Remove the pipe. Put both halves together and transfer the markings from the top of the form and of the carved area to the second half (photo 74). Mark down from the top and carve the second half.

6 Glue together as explained in step 8 on page 59. Let set and carve. See page 58 for instruction on carving polystyrene.

Other Core Materials

Andrew Goss fabricates his sculpture forms from corrugated plastic board and packing tape. This method enables him to develop his personal vocabulary of shapes in concrete without the excess weight. Look for Andrew's project on page 167.

Materials and Tools
Concrete Mix: Mix 16
Slurry Mix: Mix 21 (optional)
Corrugated plastic board
Plastic packing tape
Scissors
Chicken wire (hardware cloth or expanded metal would also work)
Wire cutters
Container to mix concrete
Rough sand paper or file

PHOTO 75

Instructions

1 Make your core form using the plastic corrugated board. (You might want to make paper forms first to help develop your ideas.) Secure sections with plastic packaging tape and seal all seams (photo 75).

2 Cover the form with metal mesh. Overlapping mesh will help to make the form stronger.

3 Work the first layer of concrete through the mesh (photo 76) until the form is completely covered. Texture this layer for better adhesion of your second layer. Cover with plastic

PHOTO 76

PHOTO 77

Faducci, *Frog on Limb,* 2002. 16 x 18 x 16 inches (40.6 x 45.7 x 40.6 cm). Cement, steel. Photo by Jim Beckett

overnight. If you can't get back to work on the piece the next day, applying slurry will help bond the second coat to the first.

4 The second layer should completely cover the first and refine the shape (photo 77). Cover with plastic overnight.

5 Sanding and filing will help further refine the form for your selected surface treatment.

Metal Armatures

The size, or scale of your project, its intended use, and the overall gesture or flow of the piece will help you to determine the material to use for your armature. Smaller sculptures, up to approximately 2 feet (61 cm) high, can be constructed from various gauges of wire, metal tubing, or other malleable metal materials. A figure with out-stretched arms needs to be constructed over an armature with a structural support that bears the weight of the gesture. As the sculpture's scale increases, it's necessary to increase the scale, strength, and durability of your selected armature materials. Armatures for larger figures can be constructed out of pipes, firmly wired rebar, bolted slotted metal, or even welded metal.

Ask yourself questions about the finished piece before building your armature. How big will it be? Is it an open or closed form? How will it stand? Is it meant for climbing? Making scaled drawings or a three-dimensional sketch, known as a *maquette,* will help you answer some of these questions and help you estimate the amount of materials you'll need.

Remember you're building your sculpture from the inside out, and you'll need to shift your thinking into 3-D. Using the human figure as an example, the torso needs to be constructed with four sides to define the front, back, and sides. If the figure is to be standing, consider using one long element that will extend from the shoulder to the foot. Using a single unit this way gives added strength to the piece that is difficult to achieve when connecting a series of smaller pieces. While the armature is like a skeleton, you don't need to duplicate it with joints for movement. In fact, you want to create a rigid construction. The same basic considerations apply to creating an abstract form or animal.

Simple Column Armature

This versatile construction technique can produce pedestals for birdbaths, fountains, or gazingballs, and with minor armature additions, even simple figures. The actual armature consists of wrapped wire mesh and chicken wire. The cardboard tube or plastic pipe is included to form a barrier that stops the concrete from filling the column. The difference in the weight of a finished column with a barrier to one without is significant. If you prefer a square base for your pedestal, you'll find those instructions on page 65.

Materials and Tools

- Concrete Mix: Premixed Sand/Topping Mix, Mortar Mix, Commercial-Grade Mason Mix, or Mix 3
- Cardboard or plastic tube
- Tape measure
- Aviator shears
- Metal mesh, ½-inch (1.3 cm) hardware cloth
- Angle wire cutters
- 6-inch (15.2 cm) pieces of 22-gauge wire
- Leather work gloves
- 1-inch (2.5cm) chicken wire
- Hammer
- Container to mix concrete
- Putty knife with tooth-shaped blade
- Plastic trash bags
- File or rasp

Instructions

1 Select the size of tube to use for the core of your armature. For most of the projects mentioned above, 4-inch-diameter (10 cm) tubes are recommended.

2 Determine the size of the piece of hardware cloth you will need. Measure the length of your tube; this will be the length of your hardware cloth. You want to be able to wrap the mesh around the tube twice with an overlapping edge (photo 78). Measure 2¼ times the circumference of your tube; this will be the width of your hardware cloth.

3 Roll the hardware cloth tightly around the tube and secure it with several pieces of wire (photo 79). If you have someone helping you with this step it will save you some frustration. One person can hold the wrapped mesh while the other twists a few the wires into place to secure it.

4 Continue to add wires every 3 inches (7.6 cm) or so to ensure that the overlapping edge lies flat.

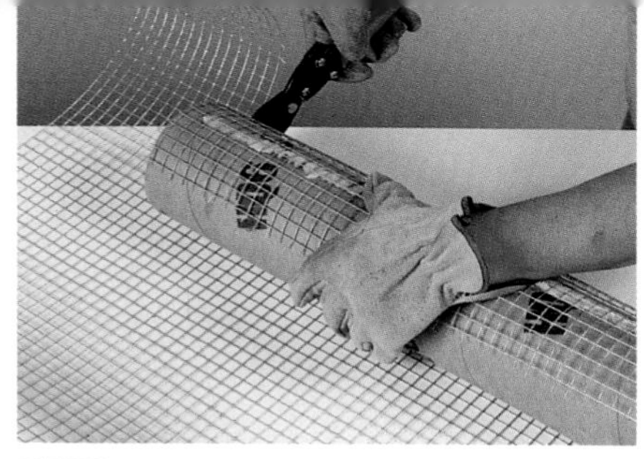

PHOTO 78

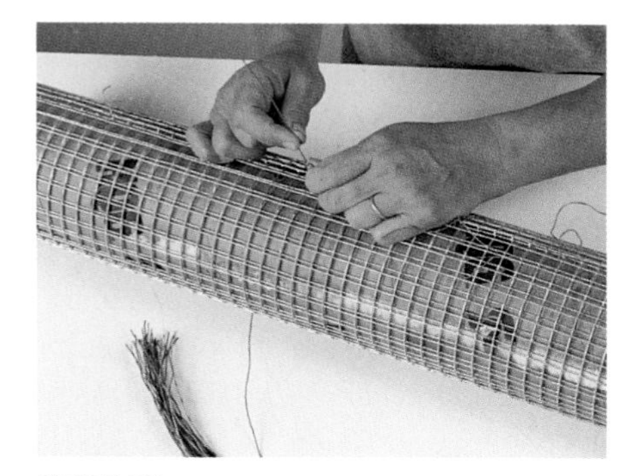

PHOTO 79

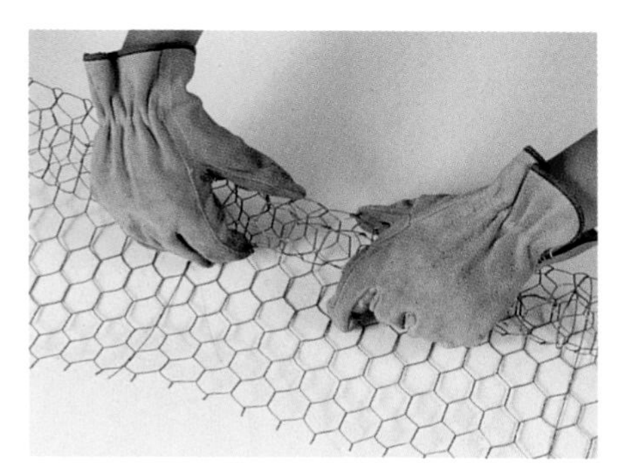

PHOTO 80

5 The foot or base of the column is the flared shape at the bottom that provides weight and area to enable the column to stand. The taller the column, the larger or heavier the base should be to counterbalance the height. The foot or base is formed with a roll of chicken wire 15 x 30 inches (38.1 x 76.2 cm). Working with the 15-inch width, roll the wire over itself to form a tube that is approximately 4 inches (10 cm) in diameter (photo 80).

6 Compress the chicken-wire tube into a donut shape by pushing the twisted element together along one side of the tube. Work it until the hole of the chicken-wire donut is the same size as the mesh-covered tube (photo 81) when one end is tucked inside the other.

7 Place it around the bottom of the tube, and use a few pieces of wire to attach it to the hardware cloth (photo 82).

8 Use the hammer to tap and form the chicken wire to prevent it from sticking up too high. Keep shaping with the hammer until it slopes down to a compressed outer edge.

9 Lay an unopened trash bag on the table or floor and place the column armature in the middle. Mix the concrete. Starting at the top of the column, apply the concrete in small handfuls, working it through the mesh as if you were grating cheese through a grater (photo 83).

10 Continue covering the hardware cloth. Since this will be your first application, some of the mesh may be visible on the surface when you finish. This is not a problem.

11 Fill the chicken wire base with concrete (photo 84). Occasionally tap the chicken wire with the hammer to insure that the concrete is being compressed through the layers of chicken wire. It's okay if some of the chicken wire shows through the concrete.

12 To facilitate additional applications of concrete, use the tooth-edged putty knife to texture the column's surface. You want to work around the column, not up and down (photo 85). Like a scratch coat in a stucco process, the ridges provide a rough surface that the next layer of concrete can grip onto.

13 When you finish, clean off any concrete that may have dropped onto the plastic bag, pull up the edges and slide another trash bag over the column so it is completely covered. Leave it to cure undisturbed until the next day.

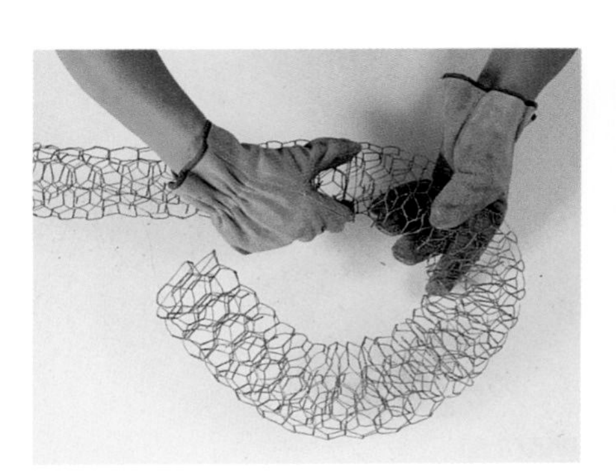

PHOTO 81

PHOTO 82

PHOTO 83

PHOTO 84

PHOTO 85

PHOTO 86

14 When you return, uncover the column. File any thin areas on the base and refine the shape as needed (photo 86). At this stage, you can continue to work on the piece, or you can spray it with water, rewrap it in the plastic bags, and let it continue to cure until you are ready to work on it again.

With the above steps, you have created a basic column. Applying more concrete will refine your form. You can also continue to develop it using various surface treatments found in Chapter 3.

Making a Square Foot for a Column

Instructions

1 Make a column armature as described on pages 63 and 64, steps 1 through 8.

2 Cut two pieces of hardware cloth to the length of the diameter of the foot by 7 inches (17.8 cm) wide (photo 87). Trim off the sharp edges to avoid being snagged while applying the concrete (photo 88).

3 Fold the hardware cloth lengthwise, and then hammer the edge to flatten (photo 89).

Additional Materials and Tools

Metal mesh, ½-inch (1.3 cm) hardware cloth

6-inch (15.2 cm) pieces of 22-gauge wire

Angle wire cutters

Tape measure

Hammer

PHOTO 87

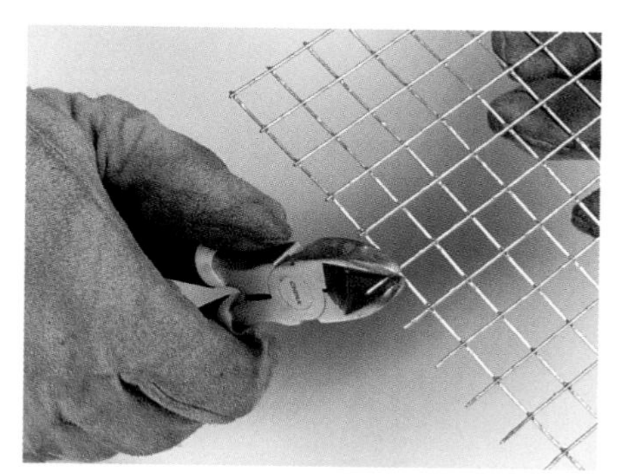

PHOTO 88

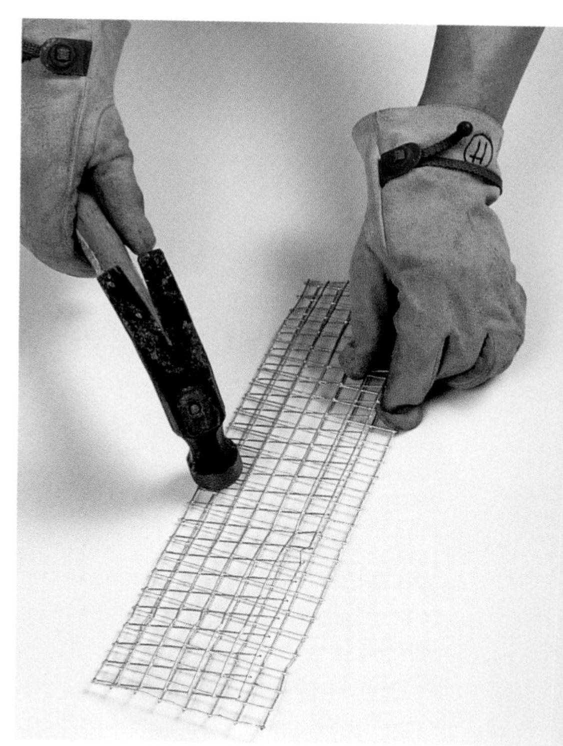

PHOTO 89

PHOTO 90

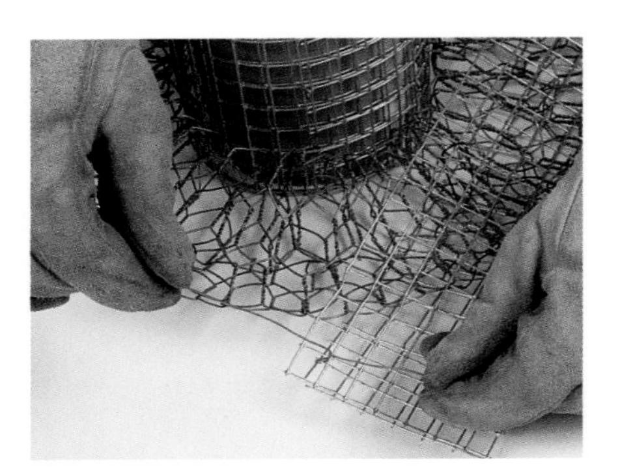

PHOTO 91

PHOTO 92

4 Slide the folded halves onto the round foot, (photo 90) and secure with a few pieces of wire (photos 91 and 92). Finish the column by following steps 9 through 14 on pags 64 and 65.

Column Connection

In the example shown in photo 83 on page 64, I have included a plastic cup at the top of the column as a mold to create a hollow sleeve for a birdbath. This column connection will fit the corresponding stem on the bowl connector on page 52. You can utilize this simple technique for other sculptural connections.

Instructions

1 Select the tube for your armature. Set, don't force, the 16-ounce (480 ml) cup into one end of the tube. If you push it into the opening, it will create wrinkles in the cup and the connecting stem won't fit into the sleeve. If it drops through the hole easily, the opening is already large enough to hold the stem connector; no cup is needed.

2 If the cup only goes partially into the hole, you need to add the length it extends when determining the length of your hardware cloth.

3 Follow steps 2 through 9 for the Simple Column Armature on pages 63 and 64. When you start to apply the concrete, take extra care to fill the space between the mesh and cup. You want to compress the concrete against the cup, but you don't want to distort the shape of the cup. The cup should remain straight in the tube opening.

4 Follow steps 10 through 14 on pages 64 and 65. In addition to refining your base, use the needle-nose pliers to remove the cup by pulling it from the rim.

Additional Materials and Tools

16-ounce (480 ml) plastic cup

Needle-nose pliers

Ricky Boscarino, *Bamboo Forest,* 1997. 3 to 12 feet (.9 to 3.7 m) in height. Concrete, glass; cast from plastic bottles. Photo by artist

Simple Animal Armature

An armature doesn't need to be complex, it just needs to bear the weight and stress of the form you want to build around it. Virginia Bullman is able to utilize a resource who's also a family member—her son-in-law is a professional welder. Virginia designs very straightforward support systems for her pieces and he welds them for her. You'll find a complete materials and tools list, plus detailed instructions for this project on page 147.

Instructions

1 Each armature is secured in a cast concrete base to provide stability and weight for the final project (photo 93).

2 Chicken wire is used to flesh-out the form and create the basic gesture, and the void in the wire form is filled with newspaper to make a core (photo 94). As the concrete is applied, the newspaper will stop the concrete from filling the form.

3 The concrete chicken is ready for its final surface treatment (photo 95). Virginia applies broken dishes and pottery (photo 96) in her trademark style to complete this feisty chicken (photo 97).

PHOTO 93

PHOTO 94

PHOTO 95

PHOTO 96

PHOTO 97

Cutting and Bending Rebar

Rebar is short for reinforcing bar and is manufactured to meet specific stress requirements. Rebar comes in eleven standardized sizes that are indicated by number. A #3 rebar is ⅜ inch (.95 cm) in diameter while a #4 is ½ inch (1.3 cm) in diameter. For most of the projects listed, we will be using one of these two sizes.

Because rebar is made for reinforcing, cutting or bending the material is no easy task. Cutting rebar is made easier with a cutoff blade on a grinder or circular saw, or by using a cutoff saw. If a hack saw is your tool at hand, cut through the rod about half way and then bend it over your knee to break the piece in two.

There is a tool designed specifically to bend and cut rebar (photo 98). This tool only makes sense if you plan on doing a lot of bending and cutting and even then, it's not that easy to operate.

Less bulky benders, like the Berkley Bender, are commercially available. However, if you just need to make a simple right-angle bend, try this: In a vise that is securely mounted, tighten your rebar at the point where you want it to bend. Select a piece of pipe that has an inside diameter that is close to the diameter of the rebar and slide it over the rod. Pull the pipe down to bend the rebar. The key to success here is having a vise that is securely anchored—once you have that, you'll be amazed at how simple bending rebar can be.

PHOTO 98

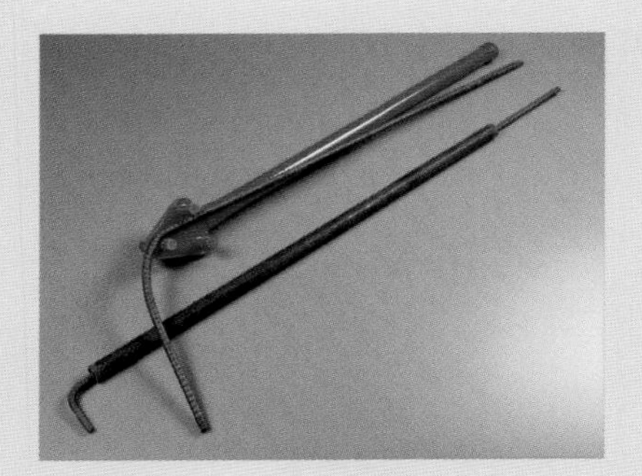

PHOTO 99

Faducci, *Cat and Frog,* 2003. 28 x 28 x 12 inches (71.1 x 71.1 x 30.5 cm). Cement, steel, glass. Photo by Solomon Bassoff

Finding a Welder

If you like the idea of having welded armatures, you can mostly likely find resources in the telephone directory under "Ornamental Metal Work," or even "Security Control Equipment." These trades weld, and would probably weld simple forms for you. I have another friend who has her welding done at a local technical school. She discusses her projects with the instructor, and he assigns them to students as school projects to train them in creative problem solving.

Simple Figurative Armature

The full instructions to complete this figure, including the cut list for the rebar, can be found on page 175. The basic information in this project can be modified as you design and create your own animals, abstracts, or other figurative pieces.

Materials and Tools

- Concrete Mix: Step 1 to 4 use Premixed Concrete, Fast-setting Concrete, or Mix 1; Step 9 use Premixed Commercial-Grade Mason Mix, Mix 3, or Mix 5
- Slurry Mix: Mix 20 or Mix 21
- 2 x 4 wood for building a base form
- Wood fasteners
- Mold release
- Mixing container for concrete
- Plastic
- Screed
- Trowel
- Pointing trowel
- Margin trowel
- Leather gloves
- Metal cutter, heavy duty
- Rebar, ½-inch (1.3 cm) diameter
- Rebar bender
- Grinder or metal file
- Lineman pliers
- Twisted tie wires (or roll of tie wire cut to comfortable work lengths)
- Wire twister
- Expanded metal mesh
- Aviator shears
- Container for mixing slurry
- Paintbrush
- Hog ring pliers (optional)
- Hog rings (optional)

Instructions

1 Make a wood frame to use as a base to provide stability for your finished sculpture. Apply the release agent to the wood. Center your frame on a plastic-covered work area.

2 Cut two pieces of ½-inch (1.3 cm) rebar the length of the height of your design plus 3 inches (7.6 cm). Bend the 3-inch portion to form an "L" (see bending rebar on page 68).

3 Mix your concrete and fill the wooden frame, tamping the concrete as you go.

4 When the frame is filled half way, position the rebar with the bent sections in the concrete. Brace the rebar so it will remain in place. (If you're using a fast-setting or accelerated mix, you'll need someone to hold the rebar for about 15 minutes.) Continue to fill the frame, screed, and trowel the surface. Allow the concrete to set overnight, or as needed, until firm enough to continue working.

5 Take your remaining pieces of rebar and begin to wire your armature together around the rebar that was set in concrete. Check your measurements to maintain the figure's proportions (photo 100). Make sure each connection is as tight as possible. I like to tie each connection with two wires placed in opposite directions.

NOTE: If your armature seems a bit wobbly, try to tighten the connection by holding on to one of the wrapped wires with your lineman pliers and give it a half turn.

6 Spray the rebar armature with primer, allow it to dry, and then fasten the expanded metal mesh to the rebar form.

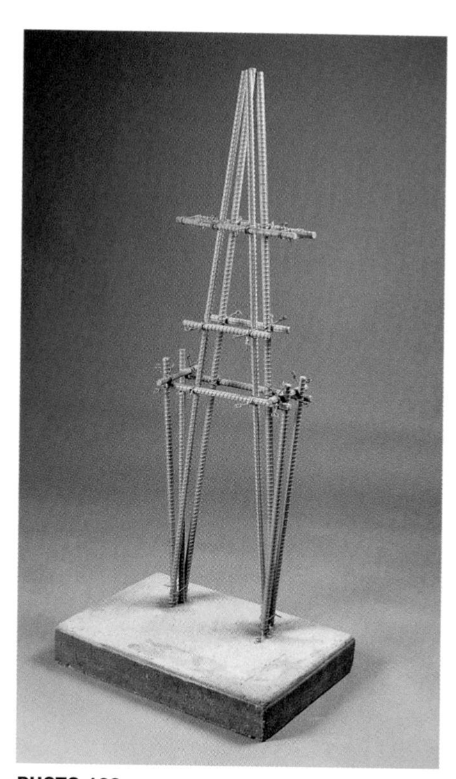

PHOTO 100

Notes on Fastening

You don't really "tie" wire, you twist it. If you're using wire off the roll, you want to first cut a comfortable length to work with. Start with an 18-inch (45.7 cm) length until you get a feel for what length works best for you. Wrap the wire snug around and behind the two rebar sections you want to connect. Using your lineman pliers, pull the wire tight, and, holding both ends of the wire together, twist until secured (photo 101).

I've come to prefer working with pre-twisted wires to make fast, easy connections. They're available in rolls or packages (see photo right). There are two tools that make their use easy and almost fun. The first tool has a hook that is set in a wooden handle that allows the hook to rotate easily. You position your looped wires around the sections you want joined, place both loops over the hook (photo 102), pull the wire ends taunt, and twist the handle quickly (photo 103). The second tool has a metal shaft over a hook spindle. to operate it, you loop the wires over the hook and pull back the shaft which causes the hook to spin (photo 104). Your piece needs to be large or well-anchored to use this tool because of the added force from pulling.

Tools for tying armatures, clockwise from top right: wire tying tool, wire tying tool, hog rings, hog-ring pliers, roll of tie wire, bundle of twisted tie wires

The tie wires work well when attaching your mesh to rebar, but there's another tool and fastener that also works great—hog rings and hog-ring pliers. You'll have to go to your local farm-supply center to find these. The brass rings come in different sizes—I try to get the large size to fit over the rebar. For attaching, you position the ring in the pliers next to the mesh and rebar (photo 105). Once you squeeze the ring on like a big staple, it will hold your mesh good and tight (photo 106).

PHOTO 101

PHOTO 102

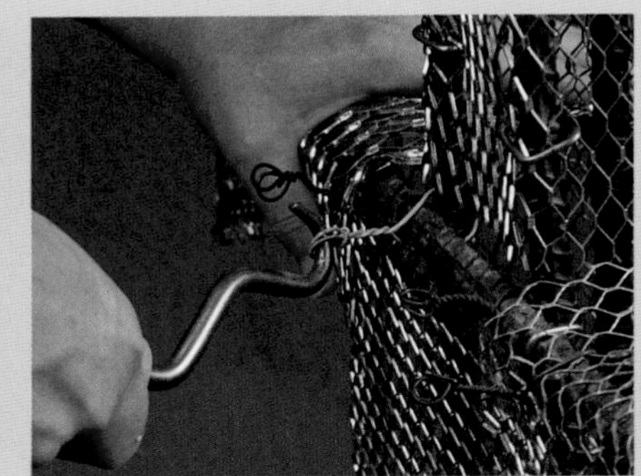

PHOTO 103

PHOTO 104

PHOTO 105

PHOTO 106

PHOTO 107

7 Additional pieces of mesh and wire can help establish the details of your figure, like the head, hair, and shoes (photo 107). At this point, I haven't yet defined the arms because I want to get a good first coat of concrete on the torso before positioning them. With that in mind, I've connected a few lengths of stainless steel wire at about waist level on the torso to help me secure them later.

8 Mix the concrete to a sticky, brownie-batter consistency, and apply it to the mesh. This isn't as easy as it sounds because I haven't utilized any additional material to create a barrier or core in this example, although that certainly could be done.

Notes on Applying Concrete

Getting the right consistency of the concrete to apply over the armature will be very helpful to your success. If the mix is too hard, you'll have to push to have it grab to the mesh, and most likely you'll end up pushing it through the mesh. If it's too soft, it will just fall off the mesh.

I like to start by using a rectangular trowel as a hawk or surface to hold the concrete supply. Using a margin trowel, cut through the concrete and slide a small amount towards the mesh. Place it on the surface by pulling the trowel in an upward motion (photo 108). The next application is handled in the same manner, but I try to calculate the placement so the second application overlaps the first (photo 109). If I push too hard with the trowel, the concrete goes through the mesh (photo 110).

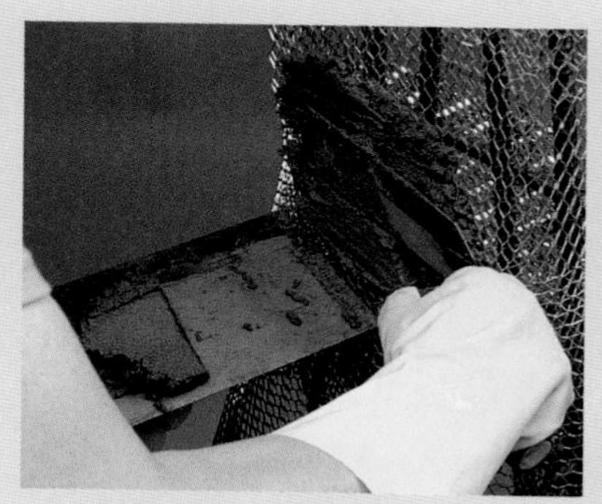

PHOTO 108

PHOTO 109

PHOTO 110

PHOTO 111

9 Not all the mesh will be covered with the first coat. That's okay. Cover the figure with plastic and let it sit undisturbed over night. The next day, uncover the figure. Remove any misplaced or stray concrete with the rub block or file. Add mesh as need to form the arms. Secure the pieces of stainless steel wire that are sticking through the torso (photo 111).

10 Mix concrete and slurry. Dampen the concrete surface, slurry, and then apply the second coat of concrete, taking care to cover any remaining mesh. While the trowel worked great for the first coat, you may opt to apply the second coat using your hands. Flesh-out the figure as needed without adding too much bulk. You will still want to model the final coat, and may also consider any decorative surface technique found in Chapter 3.

Ferro-Cement Animal

Lynn Olson uses a ferro cement process to build his sculptures. He combines networks of stainless steel rods and wires with sanded cement in such a way that the two are completely intertwined to create a strong steel and cement matrix. Lynn lent me his images of this process to use in this book.

Materials and Tools

- Concrete Mix: Mix 12 (with or without polymer), Sand Mix, Mix 13, and Mix 14
- Slurry Mix: Mix 20 or Mix 21
- Stainless steel or galvanized steel wire
- Steel fibers (steel wool)
- Wire cutters
- Needle-nose pliers
- Container to mix concrete
- Kitchen knives
- Container to mix slurry
- Paintbrush
- Rasp
- Silicon carbide paper
- Methyl methacrylate

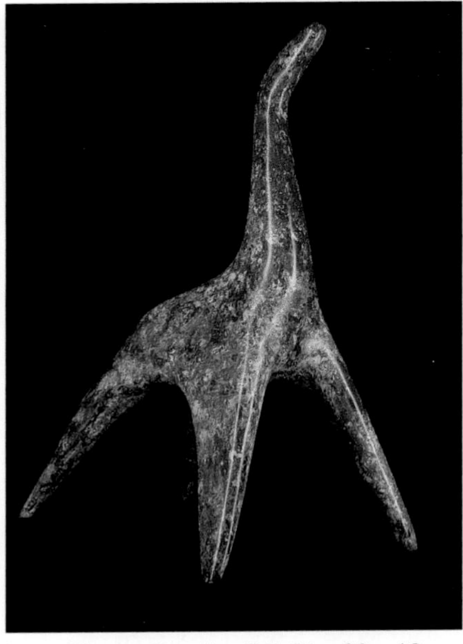

Lynn Olson, *Standing Tall.* 35 x 20 x 13 (89 x 50.8 x 33 cm). Ferro cement, sand, stainless steel rods; direct cement construction, surface filed and sanded, sealed with methyl methacrylate. Photo by artist

Instructions

1 Using heavy steel rods that are ½ inch (1.3 cm) in diameter, make a frame of your animal by first forming the legs and the body. A smaller diameter rod/wire can be used to continue developing the rest of the animal.

2 Wrap the legs with a thin gauge wire, first in one direction and then in another. Create a network by twisting the thin wire around the frame (photo 112) to define the planes of the armature.

PHOTO 112

PHOTO 113

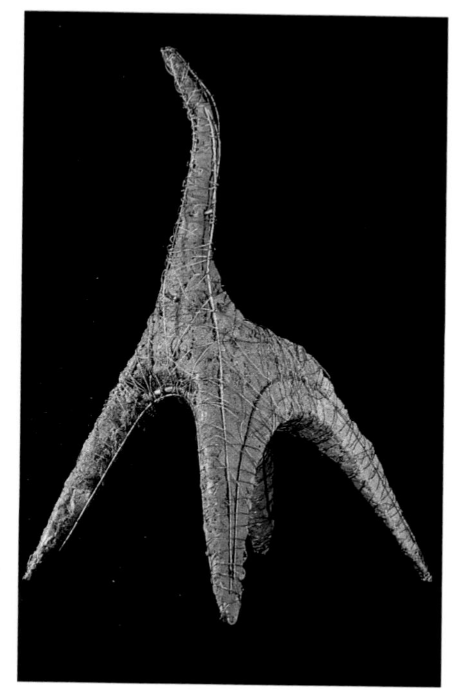

PHOTO 114

3 Mix your Basic Fiber Cement. Use a kitchen knife that has ¼ inch (6 mm) of the tip bent at a right angle (photo 113) to push the concrete into, and between, the wire network. Cover with plastic and let it sit undisturbed overnight.

4 The next day, use the coarse rasp to help define the emerging animal. On top of the first coat of concrete, add another layer of reinforcing and structural pieces. Use the heavy rod and thinner wires to create the same network as in step 2 (photo 114).

5 Mix a batch of slurry and Sand Cement. Slurry the steel (photo 115), and apply a second coat of concrete using the knife to push the materials behind and around the wires until all the wire is covered. Cover with plastic and let sit undisturbed overnight.

6 The next day, refine your animal with the coarse rasp, slurry, and then apply a thin layer of Super Fiber Cement. When the fiber cement is firmly set (6 hours to overnight), trim it with the rasp. As it hardens, finish with fine files and then finer grades of silicon carbide paper, ending with 600-grit. The surface can also be burnished to create a luster. To burnish, rub firmly and methodically, using the flat side of a steel blade or a traditional burnishing tool.

7 After the cement has thoroughly cured and dried, coat with methyl methacrylate and buff to a gloss.

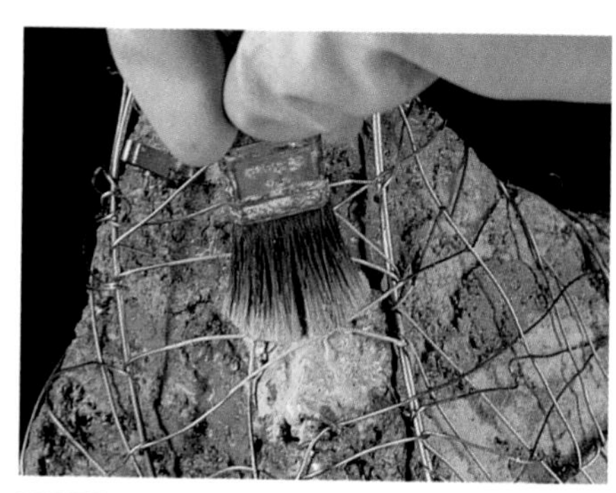

PHOTO 115

Johan Hagaman, *Birds in Breath,* 2003. 67 x 15 x 15 inches (170.2 x 38.1 x 38.1 cm). Concrete over steel, found objects, handmade objects, electric light, paint; hand formed, acid stained. Photo by artist

Direct Forming

The two main approaches to sculpting are either additive or subtractive. Carving, for example, is a subtractive process; you remove what you don't want in order to reveal the form within. Modeling, the technique of manipulating a pliable material to build up the surface, is a good example of an additive process. You'll find that concrete works beautifully for both of these approaches.

K.C. Linn, *Gladys-Agnes the Water Sprite,* 2004. 60 x 16 x 16 inches (152.4 x 40.6 x 40.6 cm). Vermiculite, Portland cement, and polyfibers over fiberglass-covered polystyrene, wire armature, PVC pipe, mirror, cement bond, acrylic paint, acrylic-based waterproofing; sponge stamped. Photo by artist

Modeling

When I talk about mixing concrete I often refer to getting a "clay-like" consistency. This is usually made from either a rich mix, with a higher ratio of cement to sand, or a mortar mix with just the right amount of water. The key to successful modeling is getting your mix to the right consistency. This is achieved when two factors are met: first, if your concrete has had enough time to fully slake—meaning the water is evenly distributed throughout the mix; and second, when there is no slump—if a mix is too wet, you'll spend more time trying to keep things from sliding than you will in adding details to your modeling. Modeling is most effective when done over a form that already has a basic coat of set concrete on it. It's not realistic to plan to concrete an armature and do the finished modeling all at the same time. Remember that working with concrete is being mindful of the stages—each process works best at a certain stage.

PHOTO 116

PHOTO 117

PHOTO 118

Materials and Tools

- Concrete Mix: Premixed Sand/Topping Mix, Mortar Mix, Commercial-Grade Mason Mix, Mix 3, or Mix 12
- Slurry Mix: Mix 20 or Mix 21
- Concrete form or concrete-covered armature
- Latex surgical gloves
- Spray bottle
- Container for mixing concrete
- Container for mixing slurry
- Paintbrush
- Sponge
- Short piece of ¼ inch (6 mm) dowel, sharpened in a pencil sharpener
- Assorted artist brushes*
- Kitchen knife
- Plastic drop cloth
- Assorted wooden or wire-end clay or sculpture tools (optional)
- Rasps or riffler rasps (optional)
- File (optional)

* I use brushes that have been worn out from painting rather than new ones. My favorites are ¼-inch (6 mm) short, flat sable, and a soft round #5 watercolor brush.

Instructions

1 Select a form that has a basic coat of concrete on it. Spray the form with water to dampen. Apply slurry with the paintbrush to the areas where you'll be adding concrete. Add one or two applications of concrete to fill out the armature or create your modeling surface. Cover with plastic and allow this to set at least 6 hours.

2 Each time you return to the piece in progress, uncover it, remove any unwanted concrete areas with the rasps or file, and spray with water till damp (photo 116).

3 Lightly slurry the area where you'll be adding concrete. Start building up the highest areas of the face using small amounts of concrete. You want to prevent air pockets from forming, so take the time to compress each application by using your fingers or a slightly damp sponge. If you're working on a vertical surface, always try to exert pressure in an upward motion.

NOTE In the beginning stages of forming, the piece will look awkward and clumsy. Don't be critical of it or get discouraged. Instead, enjoy the process as you watch the figure transform. You'll have lots of opportunities to make changes in the piece.

4 Using small applications of concrete, add material without focusing on any particular area. Establish the location or proportion of the features (photo 117).

NOTE If you notice the concrete sliding, or that there's a distortion of the modeled form, restrain yourself from continuing. Cover the piece with plastic, and allow it to setup for 30 minutes or longer before proceeding.

PHOTO 119

5 Use the sharpened dowel to help define lines or creases (photo 118 on page 75), such as the eyelids. Soften the lines by dipping the artist's brush in water, squeezing out the excess moisture, and brushing over the sharp lines (photo 119).

NOTE Think of the dowel, brushes, or any other tools you might use as extensions of your own hands. They can help to define areas by pushing and moving small amounts of concrete that would be difficult to manipulate with your fingers.

6 Repeat steps 2 through 5 until you're satisfied with your modeling. If you plan to combine a modeled element with other surface treatments, consider how much more concrete you'll need to add to incise or texture an area, or how deep a space you'll need to leave to accommodate mosaic or embedding.

Definition

Whether you're modeling or carving, definition in a sculpted form occurs in the way light and shadow play on the surface of the form. Look at the eyes in the photo on page 32, and notice how the shadow cast by the narrow ridge of the upper lids adds to the appearance of depth in the eyes. Consider how edges and shadows will add depth and the illusion of volume to your piece.

K.C. Linn, *Mr. Big,* 2002. 20 x 12 x 12 inches (50.8 x 30.5 x 30.5 cm). Peat moss, vermiculite, and cement mix over fiberglass-covered foam, oil-based gel wood stain. Photo by Elder Jones

Carving Basics

If you were carving wood or stone, you would begin by selecting a piece of that material that's close in shape to the form you want to carve. With concrete, you simply cast a form. Working with your sketches or maquette, determine the approximate dimensions for your piece and make a mold for casting. You want to cast a form that will provide enough material for developing your idea but at the same time, you don't want to cast a form that is significantly larger. Why would you want to spend all that extra energy removing material that is not part of the design?

A key aspect to enjoying this creative process is to work in a comfortable position. Position your form so both your back and arms will be in a natural position as you carve. Smaller pieces should be placed on a sturdy work surface that will both support the piece and bear any impact from the carving. If you're working on a piece that's too large to move easily, consider putting it on a turntable or lazy Susan. Because successful sculptures are interesting from all views, larger pieces need to be positioned so you can walk around them and view them from all angles as you work.

Don't be timid when you start to carve, but do work with a plan. I like to begin by marking the points that will stick out the furthest. For example, if

Sandbags

Sometimes small pieces need to be tilted to accomplish certain cuts as you carve. Using sandbags to prop and reposition a piece will help you do this, and will also help absorb any shock from the carving. A fast and easy method for creating sandbags is to take some old jeans or other sturdy-fabric trousers and cut off the legs. Sew up one end of the leg, fill it two-thirds full of sand, then sew or tie the remaining open end.

I'm carving a face, I place a mark where I want the end of the nose to be—in my enthusiasm I don't want to accidentally remove that material. Try to work systematically. If you've cast a square form but you need it rounded, start by removing the corners; it's the only way a square thing will ever get round...and don't be timid!

As you first start to remove material, use broad wide strokes. At this stage, you may be removing large amounts of concrete to define your form. As your basic form begins to emerge, you'll remove increasingly smaller amounts. Work the whole piece, developing the carving in a holistic manner to the same level of refinement the way you would develop a drawing or a painting. You don't want to get bogged down detailing a small area of the piece while the back is still exhibiting a crude start.

When you're working on an area near a nose or other protrusion, try to use your carving tools so they are going away from that part of the form. If there's no other direction you can go to carve, establish a stop line by cutting a line in front of the protrusion and then carve only to the depth of that line.

Concrete can be carved by different methods at different set stages. The tools used, energy expended, and results will vary with each stage. In fact, you might start carving a piece with a wet technique to remove a lot of material as you rough out your form and wait for a firmer set to refine smaller details. As you experiment with this versatile medium, you'll discover how to maximize its potential for your personal creative expression.

Wet Carving

The advantage of carving concrete just after it has set—when it's firm but not hard—is that the material can be removed quickly. There's also little dust as you work on your piece. The following technique uses a planter carved in the round as an example, but these same steps can be adapted to create flat forms, such as steppingstones. The tools used for this technique don't have to be fancy. You can easily collect a variety of toothed tools by picking up a package of assorted blades for a reciprocating saw at your local hardware store (see the photo below). Or, just modify some old saw blades that are lying around your shop like Elder Jones did. A visit to the kitchen area of a local thrift shop will probably provide you with a variety of knives and utensils that can be adapted.

Materials and Tools

- Concrete Mix: Premixed Sand/Topping Mix*, Mortar Mix*, Commercial-Grade Mason Mix*, or Mix 4
- Materials for constructing your casting form**
- Plastic work board
- Container to mix concrete
- Trowel
- Turntable or lazy susan
- Awl, large nails, or other sharp pointed tools
- Assorted sharp knives
- Assorted toothed and flat tools
- Whiskbroom

* If using premixed products, sift the contents of the bag before mixing with water.

**This form was made with a sheet of gutter-forming metal held together with a C-clamp and reinforced by tying a cord around the bottom.

An assortment of Elder Jones' tools for wet carving: whiskbrooms, assorted kitchen knives, nail, a variety of toothed and scraping tools made from old saw blades, many cut and shaped on a grinder

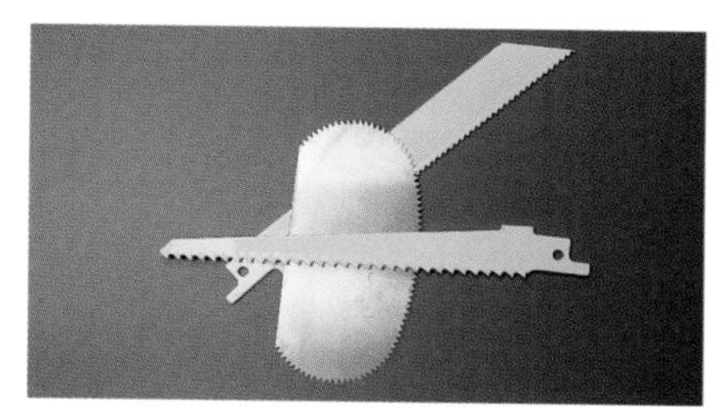

Tools for wet carving: Reciprocating saw blades and toothed ceramic rib (center)

PHOTO 120

PHOTO 121

PHOTO 122

PHOTO 123

PHOTO 124

PHOTO 125

Instructions

1 Select or build your form, and position it on the work board (photo 120). This will be the board on which you will do your carving. Working on a piece of plastic or a plastic-laminated board will provide a surface that is both non-stick and sturdy. Because this piece will be carved in the round, place the work board on a lazy Susan or turntable.

2 Mix the concrete to a creamy texture, about the consistency of muffin batter, and begin to fill the mold (photo 121). As you fill the mold, agitate or wiggle the board occasionally to eliminate air bubbles.

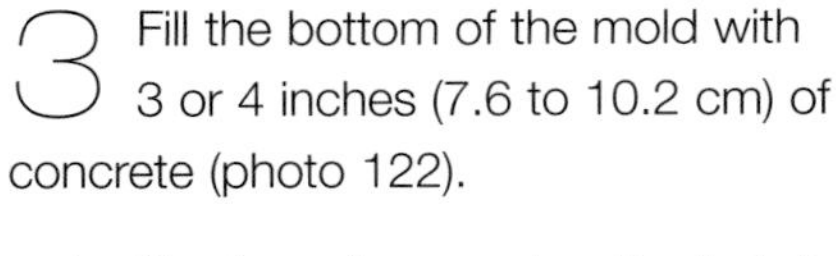

3 Fill the bottom of the mold with 3 or 4 inches (7.6 to 10.2 cm) of concrete (photo 122).

4 The inner form makes the hole in the planter. For your inner form, fill a plastic container with sand. The sand provides weight so the inner form won't float up when you add additional concrete. Place it in the concrete in the mold, then center it until there is an even amount of concrete all around (photo 123). You don't need to bury the inner form in the concrete. Continue to add concrete around the inner form to just below its lip. Tamp the concrete with a wooden block as you work to eliminate air pockets.

5 Scrape any concrete that has seeped out of the bottom of the mold off the work board (photo 124).

PHOTO 126

6 Allow the concrete to set until it's firm, but not hard. If the temperature is 70° to 75° (21° to 24°C), this will probably take about three to four hours.

7 Remove the outer form (photo 125) and then remove the inner form (photo 126).

8 Use a scraping tool to enlarge the center hole made by the inner form. Because you want to make sure that the thickness of the concrete is consistent all the way around, turn the piece as you work. A round saw blade can help you both scrape and scoop (photo 127).

9 You want to make sure that you don't scrape out too much, since it will make the bottom of the planter too thin and compromise the strength of the finished form. Check the depth by first measuring the inside (photo 128), then take a corresponding measurement on the outside (photo 129).

10 Begin carving by first defining any major sections of the form. In this case, use a sharp knife or pointed tool to incise a line for the lip of the planter (photo 130).

11 Using a large sharp knife, block out the profile shape by cutting away the concrete (photo 131).

12 Use a coarse saw blade or other toothed tool to shave and scrape as you begin to refine the shape (photo 132). As you're scraping or shaving the concrete off the form, work each stroke at a right angle to the previous stroke. You want to constantly change the direction of your carving stroke as you work across the piece to achieve a consistent surface texture.

13 Once you are satisfied with the profile of your form, use a flat-edged tool to smooth the surface (photo 133).

NOTE Don't touch the surface with your fingers. Overworking your piece or directly handling it will change the surface texture.

PHOTO 127

PHOTO 128

PHOTO 129

PHOTO 130

PHOTO 131

PHOTO 132

PHOTO 133

Elder G. Jones, *Phyto Type Planter 2,* 2002. 17½ x 16 inches (44.5 x 40.6 cm) in diameter. Concrete, tint; wet carved. Photo by artist

PHOTO 134

14 Define the lines for carving by using a sharp point of a knife (photo 134). The first time around, you want to make only a shallow incision for your lines. Each successive time around, you want to cut deeper (photo 135). Note in photo 136 how the knife blade is angled to give the cut more depth. After carving, a broad-blade tool smoothes the angles (photo 137).

PHOTO 135

PHOTO 136

15 When you're finished carving, use the whiskbroom over the entire surface of the piece to unify the surface texture (photo 138). Let the piece sit undisturbed overnight.

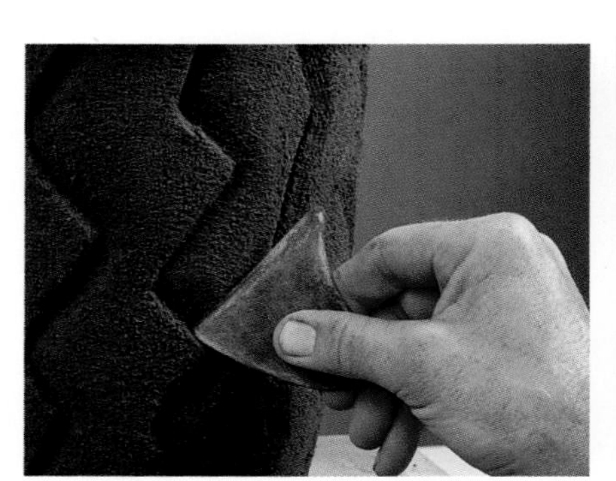

PHOTO 137

PHOTO 138

16 The next day, turn the piece over to refine the bottom edge (photo 139). Cut a hole in the center bottom for drainage (photo 140). At this point, you may want to sign and date your piece using a sharp nail. Take the piece outside and spray it with water to remove the remaining dust.

17 To cure, wrap it in a wet towel, then in plastic, and leave for five days. Or, you can leave it outside and spray it thoroughly every day for five days.

PHOTO 139

PHOTO 140

Dry Carving

Concrete is often referred to as cast stone. With this in mind, you can approach a piece of firm-set concrete much like stone for carving, using many of the same tools and safety precautions. Always wear goggles and a good particle dust mask!

The aggregate you select will have a significant effect on how the material will move and what tools you can use to move it. If you use fine aggregates like sand or powdered marble, the resulting concrete will have qualities much like limestone. Using vermiculite or perlite will produce a softer material that's easy to work. Crushed stone will make a material that is much harder and less effective to carve.

Materials and Tools

CHISELS If you're carving a block cast with vermiculite or perlite, you can use wood chisels. However, if you're carving a sand-cement block, carbide-tip stone chisels are recommended.

FILES, RASPS, AND RIFFLERS Files have cut lines at opposing directions to form edges that remove small particles. Rasps have abrasive point protrusions that remove larger particles. Rifflers are small, double-ended tools in various shapes and curves that can either have a rasp or file surface. They're very handy when working on sculptures or in tight places.

WIRE BRUSH Whenever you're using a file or rasp, you should always have a wire brush handy. It will clean the removed material from the tools' teeth and clean your tools when you're finished working.

4½-INCH (11.4 CM) POWER GRINDER WITH MASONRY OR DIAMOND-SURFACED WHEEL ATTACHMENT With the right wheel on your grinder you can effectively remove material or finish a surface. A grinder is good for carving, cutting, grinding, and polishing.

Phil Schuster, *Learning Log,* 2001. 66 x 18 x 16 inches (167.6 x 45.7 x 40.6 cm). Concrete; direct sculpted, cast. Photo by artist

Polymer Fortified Concrete and Alkali Resistant Fiberglass Mesh

Three materials are used together to create a strong, lightweight, concrete "system" that maximizes each of the materials' individual qualities. The polymer adds to the flexibility and waterproof qualities, the cement adds strength and binding properties, while the alkali resistant (AR) fiberglass scrim or mesh provides lasting reinforcing with added strength.

Materials and Tools

- Concrete Mix: Mix 12
- Slurry Mix: Mix 21
- Polystyrene foam form
- Lightweight AR glass scrim (self-adhesive or plain)
- Heavyweight AR glass scrim
- Scissors
- Permanent marker
- Container to mix concrete
- Container to mix slurry
- Paintbrush
- 4 ml plastic
- Galvanized roofing nails
- Lightweight wire
- 2 ml plastic
- Rasp or rub block
- Hammer (optional)

Instructions

1 Use the scissors to cut the lightweight AR glass scrim into strips. If I'm using rolls of self-adhesive AR tape I don't precut my strips but cut the tape as I need it. The amount and size of your strips depends on the shape of the form you want to cover. The more small curves there are, the thinner you should make the strips. Wide strips can cover large, flat areas. I usually cut stacks of 3-, and 6-inch-wide (7.6, and 15.2 cm) strips approximately 18 inches (45.7 cm) long. Any of these can be cut shorter as you work. Cutting the 6-inch strips into squares can be very useful as well. Set aside until needed.

2 Place your form on a plastic-covered work area. Mix the slurry, and slurry the entire piece (photo 141). If you're using the self-adhesive AR tape, do not slurry your piece.

3 Apply the lightweight AR glass scrim, overlapping each strip by ¾ inch (1.9 cm) or more, much in the way you would apply papier-mâché (photo 142). If using the regular AR

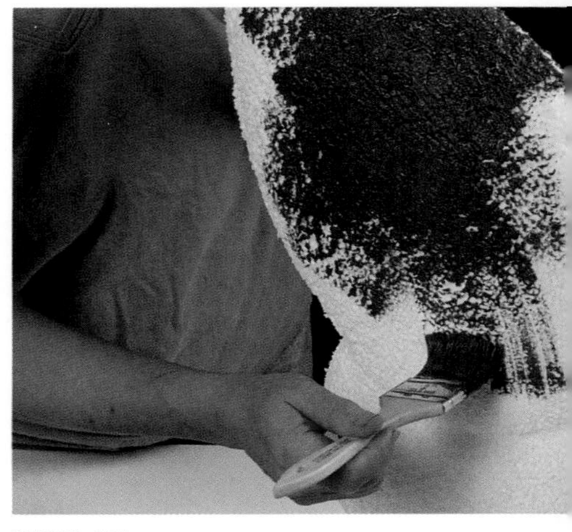

PHOTO 141

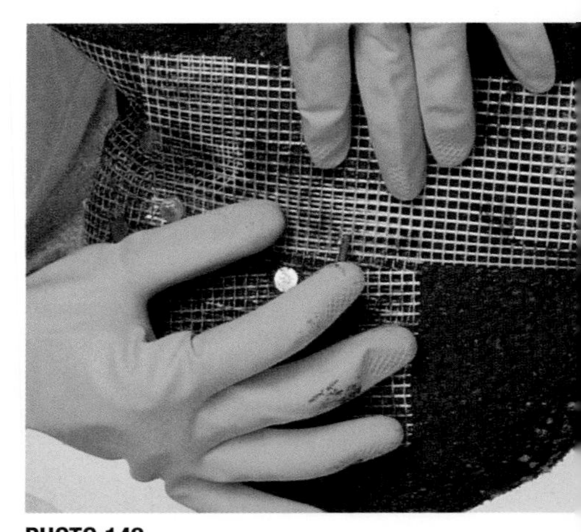

PHOTO 142

glass scrim, secure each piece tightly to the foam surface with the roofing nails. Continue until the entire form has been covered. Use as many nails as needed to ensure a secure application of mesh. Don't worry if some areas appear to have more overlapped material than others. In fact, you may want to use additional mesh to reinforce thinner areas of your form like the body of this butterfly.

4 Mix the concrete. Apply a coat of slurry over the mesh and then apply your concrete. Take small handfuls and rub it over the surface as if

you're pushing soft cheese through a cheese grater (photo 143). Make sure the concrete is worked between the layers of mesh and is compressed against the foam. Cover the entire form with concrete.

5 Use the slurry paintbrush to level the concrete by brushing lightly over the surface (photo 144). Cover with 2 ml plastic, and leave undisturbed overnight. The next day, use the rasp or rub block over the whole surface to remove any rough areas (photo 145).

6 For a small decorative piece like this butterfly, I finish it by applying another coat of slurry and cement (photo 146), and then repeat step 5. At this point, it's ready for a mosaic finish. The mosaic surface adds strength to the form. If I wanted a concrete surface, I would apply another coat of mesh with the slurry and concrete to build up the strength. If I were constructing a larger sculpture or structure, particularly if it's a piece that will be sat or played on, I proceed with the following steps.

7 Prepare your heavy scrim by cutting large sections to completely cover your shape. There are two ways to approach this. The first way would be to tack a length of mesh onto the carved shape using the roofing nails. Trace the shape using a permanent marker. When you cut out the mesh, add an additional ½ inch (1.3 cm) all the way around the tracing. An alternative method would be to tack paper over the form, and make a paper pattern that can then be traced onto the heavy mesh. Don't forget to add the ½ inch to your pattern. This will allow you to save material by arranging your pattern as conservatively as possible onto the mesh. Set the sections aside until needed.

PHOTO 143

PHOTO 144

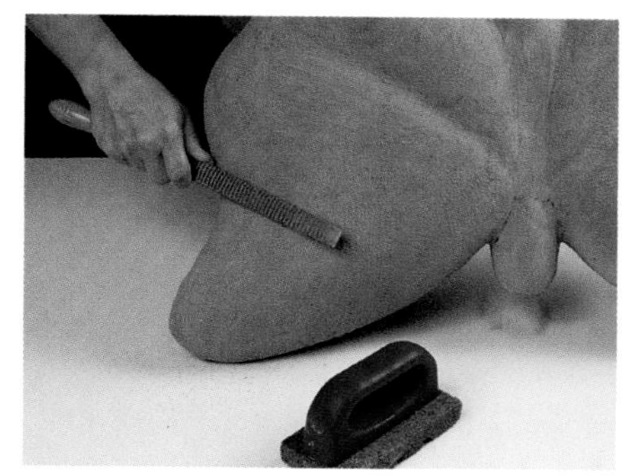

PHOTO 145

PHOTO 146

8 Mix another batch of slurry and apply it to the form. Attach the heavy mesh sections to your form with roofing nails. It may be necessary to cut slits or darts into the mesh so the mesh will conform to the contour of the form. Secure the mesh with as many nails as needed to ensure that it lies firmly against the surface. You may need to use a hammer to tap the nails through the first layer.

9 Use wire to secure the mesh sections together where the edges meet. Slurry the mesh, and then cover the entire form with concrete, using the same grating method described in step 4.

10 Even the surface as described in step 5. Cover with plastic, and leave undisturbed overnight. The next day, use the rasp or block over the whole surface to remove any rough areas. A third application of concrete will help to refine the form, give the surface a finished texture, or allow you to build up additional detail.

Simple Fountain Construction

Many of the construction methods presented could be adapted to make a fountain. This particular example has a polystyrene core construction that is covered with a polymer fortified concrete system using an AR mesh before applying the selected surface treatments. Complete instructions for this bubbling fountain can be found on page 155.

When I design a fountain, there are three basic elements that I like to keep in mind. You need a water reservoir, a place to conceal the pump, and a means to deliver the water.

The Reservoir: Sheets of foam insulation are cut to form the box that will be the reservoir. It needs to be deep enough to completely cover your pump. Glue and nail the sides together. Before covering with the polymer fortified concrete, insert a piece of pipe, slightly larger than the plug of your pump, ½ inch (1.3 cm) from the top of one of the side walls. This will help to hide your cord while also working as an overflow drain for outdoor fountains (photo 147).

Pump Conceal: A fountain is always more intriguing when you can't see the pump. I've cut my sphere in half to be able to make the space I need for the fountain workings. Draw a centerline using the cross marks on your carved sphere. Measure out an area on each half that will create a space for your pump. Use a keyhole saw to remove the sections (photo 148).

Water Delivery: Your means of delivering the water and how it will emerge (spray, gurgle, or flow) from the fountain should be determined before you even start to apply the concrete. In this example, I'm inserting a small piece of ¼-inch-diameter (6 mm) copper tubing that will extend ¼ inch past the finished surface. A section will also protrude into the pump house as a connector. A round file was used on both sides along the centerline to make a channel that will hold the copper tubing (photo 149). Once in place, glue the two halves together, tape the ends of the tubing, and begin to cover in concrete.

For the final connection, I cut a short length of plastic tubing that slides snuggly over both the copper tube and the pump connector (photo 150). I check my water flow indicator on the pump and fill the reservoir with water before plugging in the pump. Then, I sit back and enjoy!

NOTE: To carve a sphere, I first start with a cube with sides equal to the desired sphere diameter. Draw a diagonal line, corner to corner, on each face of the cube. The outer edges of the sphere are where the lines cross. Start by removing the corners using a handsaw. Each time you saw off a corner, you create new ones, although they're at a lower profile. Just keep removing corners. Do not remove the material around the "Xs." When the form is sufficiently roughed out, switch to your wire brush and continue to remove material from between the "Xs," constantly turning the form until you have a sphere.

PHOTO 147

PHOTO 148

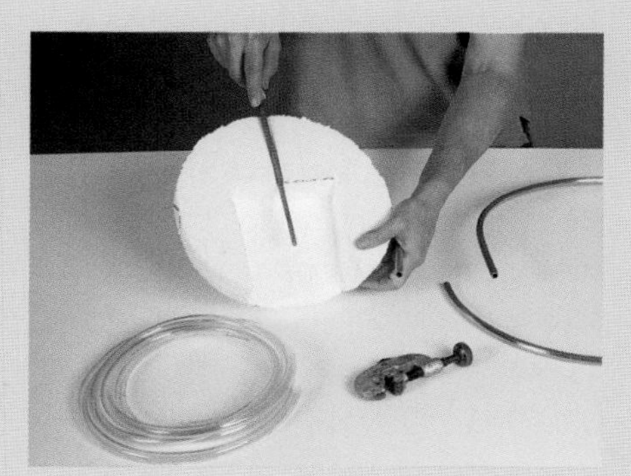

PHOTO 149

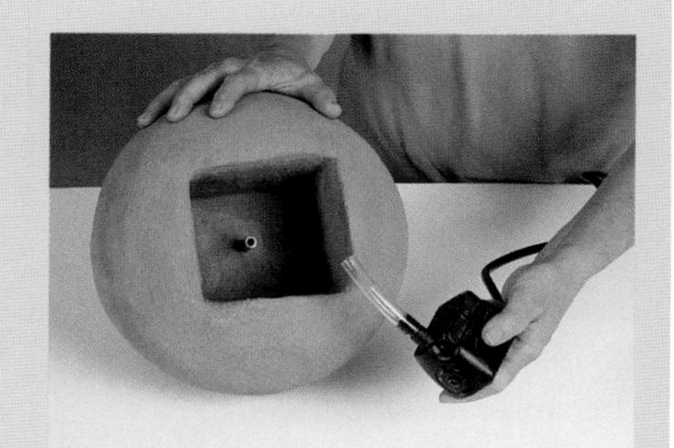

PHOTO 150

Upper left:
Phil Schuster, *Untitled*, 1998. 30 x 2 ½ x 3 inches (76.2 x 6.4 x 7.6 cm). Fiber concrete, polystyrene foam, paint; direct sculpted. Photo by artist

Lower left:
Faducci, *Iguana,* 2002. 16 x 66 x 26 inches (40.6 x 167.6 x 66 cm). Cement, steel, glass. Photo by Jim Beckett

Above:
Faducci, *Dog with Wings*, 2003. 29 x 21 x 12 inches (73.7 x 53.3 x 30.5 cm). Cement, steel, glass. Photo by Solomon Bassoff

CHAPTER THREE

Surface Treatments

The natural color and surface of concrete has its own raw beauty, but sometimes you may want a little more finish. Today, the possibilities for manipulating the surface, color, and texture of concrete have never been so wide reaching. Color can be added or applied, mosaic materials can be adhered or embedded, and surfaces can be stamped or polished.

Many finishes are achieved with your final application of concrete, some during various set stages, while others are applied only after your concrete has cured to avoid blistering or efflorescence. Experimenting with these techniques may help you develop surface treatments and finishes of your own.

Textured Surfaces

We're used to seeing the smooth-troweled floors of warehouse stores, the brushed surfaces of sidewalks, and many variations of mortar applications between bricks. These seemingly mundane surface treatments not only reveal the tools of the tradesmen, but also their sensibility to the craft. You can manipulate the surface texture of concrete at every set stage starting with the initial set to the hard-cured surface. The firmness of your material determines what type of mark or finish you can apply.

In addition to learning about the surface treatments in this book, looking into standard construction methods can expand your vocabulary of tools and techniques. Professionals who pour patios and driveways have all sorts of tools and methods to create textures and patterns. Stucco and plaster finishers can provide insights on the application of wet materials to create decorative designs, while specialists making concrete countertops can provide you with information on building molds and polishing concrete.

Exposed Aggregate

Aggregate, while a key component in your concrete mix, can be a thing of beauty in and of itself. That's very evident when you look at the photo on page 11 and the photo on the right. Sometimes the aggregate is so beautiful that you want it to remain exposed—and it all has to do with timing. This process works best on horizontal surfaces like bench tops or stepping-stones. Exposed aggregate is also referred to as seeding. With this method, you'll get the color of the decorative aggregate as well as an interesting surface texture in your project.

Decorative aggregates, clockwise from top center: amazonite, leopard-skin rock, mother of pearl, tumbled stain glass

Materials and Tools

- Concrete Mix: Premixed Topping/Sand Mix, Mortar Mix, or Commercial-Grade Mason Mix, or Mix 3 or Mix 2*
- Mold
- Screed
- Float
- Decorative aggregate
- Spray bottle or spray attachment for hose
- Wire brush

*If you use Mix 2, skip steps 1 and 2 and start with step 3.

Instructions

1 Cast your mold. Level the mold with your screed, and allow any bleed water to evaporate.

2 Wash your aggregate. Then spread the clean aggregate in a single, even layer over the entire surface.

3 Tamp the aggregate into the surface with your float, and then use the float over the surface until a thin layer of concrete covers the aggregate. Don't use the float excessively; you don't want to work the aggregate in too deeply.

4 Let the concrete set for 30 to 60 minutes. Mist a small area, and then use a wire brush to scrub the surface. If the material comes loose, tamp it back in with the float and wait about 30 minutes more. Don't wait too long or the concrete will be difficult to remove. When you can work on the surface without removing stones, mist and scrub the entire area. Rinse well.

5 Cure concrete for one week. Clean again with water and wire brush. Use a weak solution of muriatic acid to remove any remaining residue (see pages 25 to 27). In cases where the concrete has gotten away from you and is too hard to scrub off, you can recover some of your aggregate color by using a grinder to remove a layer of concrete. Add a topical sealer, and your surface will slightly resemble polished concrete, having achieved its sheen from the coating rather than through polishing. Concrete can also be sand-blasted, a technique frequently used on the vertical surfaces of buildings to provide an exposed aggregate finish.

Incising

As a child, did you ever write your name into a freshly poured concrete sidewalk? If you did, then you already have experience with this technique (sort of). Incising is the process of drawing into the surface of semi-set concrete with a sharp instrument. You can incise designs, words, or your signature into your concrete creations. Use this technique on the final application of concrete to a base form or to decorate the surface of a casting. It works well on both horizontal and vertical surfaces.

Incising on the Surface of a Base Form

Materials and Tools

Concrete Mix: Premixed Sack Sand/Topping Mix, Mortar Mix, Commercial-Grade Mason Mix, or Mix 3

Slurry Mix: Mix 20 or Mix 21

Container to mix concrete

Container to mix slurry

Paintbrush

Incising tool(s)*

Artist's paintbrush

*Tools can be a pencil (I find a dull one works better than a freshly sharpened one), a short piece of ¼-inch (6 mm) wood dowel that's been sharpened in a sharpener, a nail, or an awl—whatever feels comfortable in your hand. Experiment with other mark-making tools. A bent piece of pipe can make a clean deep groove, while a broken piece of comb can make a rough, scratchy surface.

Instructions

1 Mix your slurry and concrete. Dampen the base form with water.

2 Brush the slurry onto the base form, and apply a coat of concrete at least ¼ inch (6 mm) thick. Work the surface of your piece until you're satisfied with the overall texture. Allow the concrete to set until firm, approximately 30 minutes.

3 Select the tool(s) you want to use to make your incised design. Sketch the design or words very lightly in the surface. If you need to make some changes in how the design is positioned, refinish the concrete surface and try again.

4 Once you're satisfied with the sketched design, use your tool so that it's at about a 45° angle to the concrete surface. You're not just writing or drawing, but pressing the tool into the concrete. Don't make your marks too deep on the first pass.

5 Retrace your lines, pressing slightly deeper so that the finished depth of the line is approximately ¼ inch (6 mm), or at a level that looks good to you.

6 Use the artist's paintbrush to soften the lines. Dip the brush in water, squeeze out the excess moisture, and brush lightly over the lines. Rinse and repeat as needed.

Marvin and Lilli Ann Killen Rosenberg, *Loveseat,* 2003. 3 x 5 x 2½ feet (0.9 x 1.5 x 0.7 m). Pigmented concrete, handmade ceramic inserts, Venetian glass; welded armature with wire mesh and insulation foam, mosaic. Photo by artists
Location: Eugene Central Library, Eugene, OR

Incising on the Surface of a Cast Form

Materials and Tools

Concrete Mix: Premixed Sand/Topping Mix, Mortar Mix, Professional-Grade Mason Mix, or Mix 3

Mold

Everything listed in Materials and Tools for Incising on the Surface of a Base Form, except for the container to mix slurry.

Instructions

1 Cast your mold. Screed and/or trowel the top surface to your desired smoothness.

2 Allow the bleed water to evaporate and the surface to become firm to the touch.

3 Follow steps 3 and 4 from above. If your lines fill with water, let the surface moisture evaporate longer before proceeding.

4 Follow steps 5 and 6 from above.

Stamping

Use old printing blocks, rubber stamps, cookie cutters, various sizes of tin cans, and even buttons, shells, and plastic toys to press into the surface of semi-set concrete. Check the kitchen for implements, such as cookie presses or potato mashers, that will add interesting textural elements. Simply dip the "stamp" in water, shake off the excess moisture, position it on the surface, and press to your desired depth.

In case you're thinking about stamping a much larger area, say a patio, commercial stamping tools are available at most large rental centers that can imprint simulated paving brick, stone, tile, or other patterns.

Textures in Fresh Concrete

As you apply concrete to a form either by hand or with a tool, you'll notice a unique impression or mark that's the result of that application. As the application continues and the mark is repeated, a texture begins to emerge. Trowels, palette knives, table knives, and small putty knives are all going to produce different textures, depending on how you stroke or apply the wet concrete. Distinctive marks made when you apply pinches of concrete by hand or make dents with your fingers in a modeled surface will add to the personality of your work. Each tool and each person leaves its own unique set of marks, so the possibilities of new and interesting textures are virtually limitless.

HAND MARKS

I've talked about concrete with a clay-like consistency. One of the reasons people respond to many clay objects is that they hold the finger marks of the person who made them. Your concrete can also hold the "maker's mark" when you use your hands to create a surface texture.

Stamping tools, clockwise from top right: vegetable masher, metal cans, found metal objects, wooden dowel sharpened in pencil sharpener, artist's brush, plastic letters, old gears, crumpled aluminum foil, (blue) flexible molds made from rocky surfaces

Texturing A Mask

Materials and Tools

- Concrete Mix: Premixed Sand/Topping Mix, Commercial-Grade Mason Mix, Mix 3, or Mix 4
- Slurry Mix: Mix 20 or Mix 21
- Premade mask form
- Masonry drill bits
- Drill
- 4-inch (10 cm) grinder with masonry wheel
- Container to mix concrete
- Container to mix slurry
- Paintbrush
- Incising tools
- File or wire brush

Instructions

1 Start with a pre-made mask form (pages 48 through 51). Mark the locations for the eyes and the mouth.

2 Use a large masonry drill bit to drill through the concrete where you have marked the eyes (photo 1). Let the drill bit do the work. You don't want to press too hard, particularly if the concrete mask is not fully cured.

3 Use a masonry drill bit to drill holes at the corners of the mouth. Use your 4-inch (10 cm) grinder with the masonry wheel on it to cut out the concrete section between the holes of the mouth (photo 2).

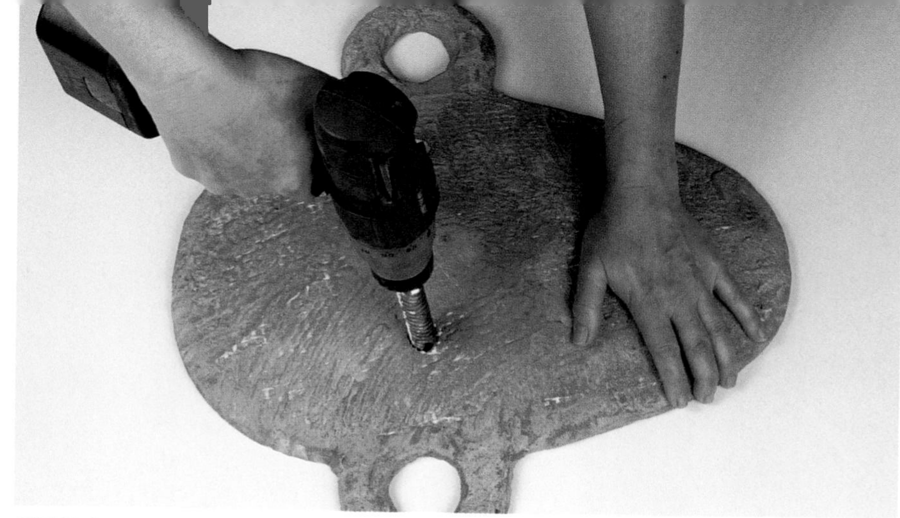

PHOTO 1

PHOTO 2

PHOTO 3

PHOTO 4

PHOTO 5

4 Mix your concrete and slurry. Mist your mask form with water and paint on the slurry. Add concrete to the mask form to cover the entire surface and to build up features such as the nose (photo 3).

5 Draw your fingertips over the surface to begin to make marks in the fresh concrete. Heavy lines on the forehead, circles around the eyeholes and lines on the cheeks will form stylized expressions reminiscent of tribal cultures. Use incising tools to help define your marks (photo 4). Be spontaneous. Don't worry if small balls of concrete appear as a result of your texturing. These can be removed the next day with a wire brush or a file.

6 Cover with plastic until the next day (photo 5).

TOOL MARKS IN HARDENING CONCRETE

As we've seen in Carving Basics in Chapter 2, set concrete can be cut with a knife, and an emerging form can be developed with small pieces of old saw blades (see the photo below). A coarse rasp cut across a surface creates a series of small lines that serve not only to form your piece but to add surface interest as well. There is another type of rasp that has sharper holes like a cheese grater that can slice through soft concrete, leaving flatter but still distinctive lines.

As concrete continues to harden into a rock-like mass, stone-carving tools, such as flat or toothed chisels and points, can be used to create still other textures. This effect can be heightened in a piece by contrasting different surface textures to accentuate the forms of your design; incorporate smooth restful areas to set off busy textured ones.

Incising into set concrete

Polishing tools, clockwise from top right: wax, buffing pad, 6-inch (15. 2 cm) diamond pads, diamond hand-sanding pads with foam backing, waterproof sandpaper, hand sander with hook-and-loop backing, assorted grits of sanding pads with hook-and-loop backing (color coded), right-angle grinder/polisher with built-in hose connection and ground fault plug, squeegee 4 ½-inch (11.4 cm) diamond grinding and polishing pads (graded grits are color coded)

Polishing by hand

Polishing Concrete

Polished concrete has become an intriguing surface solution for low maintenance floors, countertops, and architectural details. Artistically, sculptor Lynn Olson (see Three Arches on page 172) has inspired generations of artists to see concrete in a completely different way. His polished, ferro-cement forms look more like marble than concrete.

Like most surface effects, there is more than one way to achieve a polished or polished-looking finish, however the basic approach is the same. Whether you are sanding wood, or polishing stone or concrete, you need to start with a coarse-grit abrasive and work your way through successively finer grits until you reach the desired finish. Abrasive grits are categorized by numbers: the smaller the number, the coarser the grit. Diamond is the abrasive material used for polishing concrete. Most concrete polishing starts with a grit number below 100. I use 60 grit and then work through 150, 300, 500, to 1000. Discs with various grits are available for grinders, and pads are available for handwork. The process is really divided into two different stages: first the grinding and then the polishing. There are no published standards for what constitutes polished concrete, although some purists believe that you need to finish with grits as fine as 1800 to 3500 to be truly polished. Even without going that far, you can achieve beautiful results.

When using power equipment, you can also choose whether to polish wet or dry. The water used in the wet process helps to keep the abrasive discs cool and lubricated, extending the life of the disc. It also produces a slurry residue as the material is ground and mixed with water. Dry grinding produces a lot of dust that must be collected. Any sanding, grinding, or polishing technique will require that you plan your work area and take the necessary safety precautions.

Anything that can be done with a power tool can also be done by hand, as shown in the photo above. Sanding pads and palm sanders are available with hook-and-loop backs. The pads can also be used independently over curved areas or in tight spots.

Wet Polishing with a Grinder

When looking at options for wet grinding, I opted to use a grinder that was made for the job rather then trying to have someone supply the water or use an adapter kit. I like to be careful when mixing electric power tools and water. The grinder pictured in the photo on page 92 has a hose connector and water is dispensed from the center of the disc.

Materials and Tools

- Concrete Mix: Premixed Concrete or Mix 2
- Cast concrete slab
- Water grinder
- Rigid backer pad (backed with hook-and-loop tape)
- Assorted grit pads 60, 150, 300, 500, 1000 (backed with hook-and-loop tape)
- Skid-proof mat
- Squeegee
- Rubber apron (optional—but know you'll get wet)
- Rubber-soled shoes

Instructions

1 Place the slab on a solid work surface. Use a skid-proof mat for smaller objects.

2 Attach a 60-grit disc on the grinder. Plug into a grounded fault plug if one is not provided as part of the grinder. Attach the hose and check water flow.

3 Turn on the grinder and water before placing the grinding disc onto the slab (photo 6). Keep the disc as flat as possible on the concrete surface to avoid gouges or waves. Work at a uniform speed back and forth across the surface.

4 After grinding off the initial layer of concrete, turn off the grinder and water and use the squeegee to remove excess slurry and debris from the surface (photo 7).

5 Resume grinding as described, but at a right angle to your initial passes. Use the squeegee to clean the surface.

6 After two more passes, or once you have achieved a uniform look across the surface, turn off the water and the grinder and change discs to the next higher number.

7 Continue working through the higher numbered discs until you reach the desired surface.

PHOTO 6

True Grit

Professionals who polish concrete regularly know by the look of the surface when to change discs. Each person will have their own timing depending on the type of equipment they use and how they use it, how fresh the discs are, and the cure level of the concrete. There is no way for me to provide you with exact information on how many times you need to go over a surface with a specific grit. You'll have to experiment and trust your own sensibilities as you work. Just remember that there is no shortcut. You really do need to work through the various degrees of abrasives as you remove successively smaller scratches to produce a polished surface.

PHOTO 7

Cement Paste Backfill

If you've worked the surface with your 500-grit disc but still think the surface feels rougher than it should, it might be the result of air bubbles that were not vibrated out of your initial casting. No matter how much you polish, you will still feel the texture of those small pits. This process can solve that problem and can also be utilized as a decorative surface technique (see Spike Sculpture on page 167).

PHOTO 8

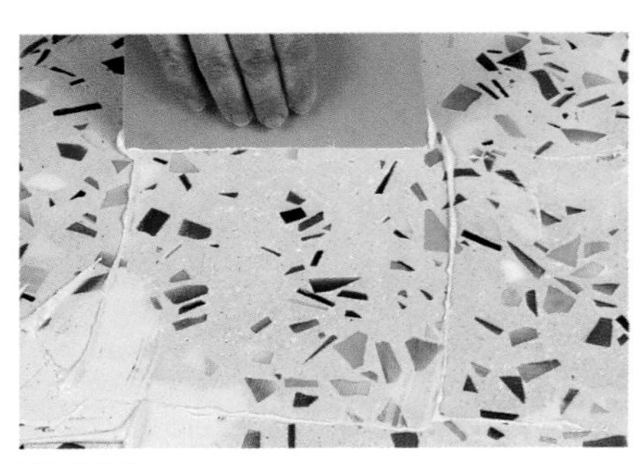

PHOTO 9

PHOTO 10

Materials and Tools

Concrete Mix: Mix 15 or Mix 19

Small mixing container

Plastic spreader

Instructions

1 Mix up your cement paste (photo 8). Use the plastic spreader to firmly press the paste across the surface (photo 9).

2 Work at right angles for each application to insure full coverage. Remove excess paste with the spreader.

3 Resume polishing with your higher numbered discs. When you're finished you should notice an improved surface texture (photo 10).

Mosaic

One of the things that attracted me to concrete was that I could make large sculptural forms that could go outside. I found that these forms were perfect for covering with colorful mosaic. The aspect of mosaic that will be addressed in this section is choosing the right materials to create weather-resistant surfaces. By selecting the correct materials and techniques, mosaic can provide a long-lasting, low-maintenance, colorful surface for exterior concrete work.

There are three variations on the mosaic process that will be covered. The first two, the direct and indirect methods, are more traditional. With the direct method you take a piece of mosaic material, apply adhesive to it, and press it directly on a base surface. With the indirect method you develop your design on one surface and then transfer the mosaic design onto another surface. The third process is embedding. Here you push the mosaic pieces into a bed of wet concrete. It's more difficult to get fine detail with embedding than it is with direct or indirect mosaic techniques, but it's a more spontaneous method of working. The information for making direct and indirect mosaic is very similar. There's a separate section on the embedding process on page 106.

To mosaic a concrete base form, it must be cured and dried before starting. The mosaic materials are attached to the surface of the base form with adhesive. Each piece of mosaic should have a space, or *interstice,* between it and the next one;

Marvin and Lilli Ann Killen Rosenberg, *Lizard Bench,* 2003. 18 x 60 x 14 inches (45.7 x 152.4 x 35.6 cm). Pigmented concrete, handmade ceramic inserts, powder coated welded steel frame; mosaic. Photo by artists. Location: Eugene Central Library, Eugene, OR

they should not touch. Once the adhesive has dried, a cementitious material, commonly known as grout, is pushed into the spaces between the mosaic pieces. The grout is leveled to and then cleaned off the mosaic materials to create a solid durable surface.

There are additional considerations when creating an outdoor mosaic piece—mainly the weather. The biggest enemy is the freeze/thaw cycle. Not all climates need to be concerned by this, but if you live in an area with changing seasons, they can wreak havoc on a poorly constructed piece or one made from the wrong materials.

If you live in an area that experiences frost, ice, or snow, you'll want to be careful when selecting materials. As you accumulate your supplies for outdoor pieces you need to look for frost-proof or frost-resistant materials. When I shop at my local home improvement store for appropriate outdoor materials, I refrain from explaining my current creation—it confuses most people—so I just ask a simple question, "Will this material (tile, adhesive, grout) work for my swimming pool?" I don't have a swimming pool, but the concerns of moisture and year-round exposure to the elements are the same, and I get a quick answer instead of a puzzled look.

Safety Glass

Recently, I led a mosaic project with young children for an organization that didn't have a budget for supplies. I'd been given large buckets of stained glass from friends but knew that the glass would be too sharp for kids to use. I just put them into my concrete mixer with sand and water and let them tumble for about 2 to 3 hours. That was all the time needed to make a safe, free, recycled mosaic material. The children had a great time, and no one got cut!

Mosaic Materials

CERAMIC TILE The basic bathroom tile that's most commonly available is not suitable for outdoor use except in frost-free climates. This tile is made of a porous, low-fire clay body that will absorb moisture. When the temperature drops, the freezing water will expand, cracking the tile or causing the glaze to flake off the surface. If you don't live in a frost-free climate, you will need to collect frost-proof or frost-resistant tiles. Most manufactures will only rate their tiles frost-resistant to avoid any consumer problems. High-fired stoneware and porcelain are the two types of clay most often used for exterior tiles, although some manufacturers will carry terra-cotta tiles that are frost-resistant. When in doubt, ask.

You will also notice that some tiles are glazed and some are not. Glaze is a suspension of powdered glass that has been melted on the surface of the tile during the firing process. Glazes can be either glossy and shiny or dull and matte. The amount of shine on the surface doesn't factor into whether or not the tile can be used outside. There is still another type of tile that has color mixed into the clay itself so that the color is integral to the clay body. This type of tile is porcelain and works well for exterior pieces.

Vitreous is another key word when looking for exterior mosaic materials. When the material is kiln fired to a high enough temperature to make the molecules of the material move as close together as possible, it becomes vitrified. Since vitrified materials are not porous, they will not take on moisture and therefore can't be affected by the freeze/thaw cycle.

Ricky Boscarino, *Undulating Wall,* 1998-2001. 55 feet (16.8 m) long, 11 feet (3.4 m) maximum height. Portland cement, diamond-wire lathe, high-fired ceramics, glass, pebbles; mold-cast concrete, ferro-cement form, mosaic. Photo by artist

GLASS By nature, glass is a vitreous material and is ideal to use. Traditional glass mosaic materials include smalti, small glass rectangles that are hand cut from poured cakes of glass, and cast ¾-inch (1.9 cm) squares referred to as vitreous glass. Both are beautiful materials that are commercially available but can be costly for large-scale projects. Stained glass, also suited for mosaic applications, is more readily available. You can purchase stained glass at hobby centers or stained-glass studios. You might even know a stained-glass artist who wants to get rid of their scraps. The only problem I've had with stained glass is that larger glass sections sometimes crack with the minimal expansion and contraction of concrete during the freeze/thaw cycle. I haven't seen this occur when I've used smaller pieces. Marbles, glass gems, beads and found glass, such as sea glass, broken stemware, or colored bottles can also be used as interesting surfaces or accents.

MIRROR You can get some great effects by incorporating mirror in your mosaics. Rather than using thin, low-grade material, try to use high-grade commercial mirror that is about ¼ inch thick (6 mm). The caustic chemicals in cement-based products will, over time, alter or damage unprotected mirrored surfaces, causing dull gray spots to appear. I like the natural antiqued look, but you can extend the life of your mirrored sections by first applying a good coat of spray acrylic sealer onto the back of your mirror. Spray in one direction and, after that's dried, spray at a right angle to the first coat to insure a good even coverage.

BROKEN DISHES In selecting dishes to use in your outdoor piece, follow the same advice as with ceramic tile: high-fired and vitrified. A popular, reissued 1940s dinnerware works well because it's made from porcelain. Grandma's china may be porcelain, but you'll have to check carefully. In addition to checking the markings on the back of plates, I have a test that I use in the privacy of my studio. I only mention it here as an aside; it certainly isn't scientific. I touch the clay edge of a piece of broken dinnerware to my tongue. If it sticks, I assume that it's a porous low-fire clay bodied dish only suitable for interior or warm climate work. If my tongue doesn't stick, I'm comfortable incorporating it into an exterior piece. We've all been told by our mothers not to put strange things into our mouths, so you decide if you want to use this test or not.

ADDITIONAL MATERIALS Polished rocks, stones, shells, nonferrous metal objects like brass keys, copper pieces, or silver can also be used. When using broken jewelry, be cautioned that plating will quickly disappear. In many cases, the chemicals in the cement will cause a reaction—changing the metals' color and causing them to tarnish—creating a new patina on the surface. Ferrous metals can also be used. Kem Alexander uses ferrous metal objects, and their resulting rust, as a signature design element (see photo above). You can find out more about using metal objects and rust as a colorant on page 114. You can also use wood and plastic objects, as long as you understand that they will not be very durable.

Mosaic Tools

Mosaic materials can be smashed with a hammer to make random pieces, or cut with a specialized saw to make precise geometric shapes. You'll find that your sense of style and design will start to come into play as you prepare your materials.

SAFETY GOGGLES, GLOVES, AND EAR PROTECTION Protect yourself when preparing your mosaic materials. Safety first!

HAMMER Ceramic tiles and plates can be safely broken with a hammer by placing them between sections of newspaper. Place them with the glaze side down to minimize glaze chipping, and clear the newspaper of scraps and chips between each group you hammer. A few well-placed hits will give you a good assortment of shapes. Avoid pulverizing useable material. Hammering glass will result in slivers rather than nice shapes. Use a cutter for glass.

TILE NIPPERS Nippers can be used to modify larger pieces of tile or plates and sometimes glass. They can be found in the tile section of your home improvement center or at a hobby store that sells mosaic supplies. When you look at your nippers, you'll notice that the jaws do not close. They are not meant to be a cutting tool like scissors, but are designed to encourage a crack that will break your tile. You'll have more success using nippers if you just put a small amount of the tile into the jaws of the nippers (see photo 11). Less (tile in the nipper jaw) is more (success when cutting tile). You'll also see a curve or a hook in the jaws of the nipper. Position the curve so it's facing toward the center of your body. Hold the nippers at the back of the handles; don't choke up on them. Place only ⅛ to ¼ inch (3 to 6 mm) of the tile in the jaws of the nippers. You'll also notice that the end of the nippers is flat. Adjust the flat surface so that it's pointing in the direction you'd like the tile to break. Squeeze the handles firmly, using one or both hands as needed. Voilà! If you need a little extra leverage, try holding the handles against your thigh or the edge of a table as you squeeze. Tile nippers can also be used as a shaping tool to nibble away small sections of a plate or tile to define a shape. You can remove material, using either side of the jaw, by moving in small increments to form exterior curves. Be patient. This tool takes practice and some hand strength.

TILE CUTTERS This tool combines two tools that are often used with stained glass. It has a carbide wheel for cutting and a pair of flange-like wings at the end that are similar to running pliers. The key thing to remember about this tool is that it's just for cutting, not for trimming. When you're cutting a piece in two, each section must be at least as wide as the flanges to be successful. Hold the tool handles together and rest the wheel on an edge of the tile surface. Press down firmly as you roll the wheel across the tile to the other side. This action will score or cut a line into the tile surface. I do this by holding the handles together with one hand and placing the other hand on top of the tool. You want to score the line with a single pass. Do not roll the wheel back and forth or try to score over the same line. If you're using the tool correctly, it will make a sound as annoying as someone scratching fingernails across a blackboard. Once the line is scored, open the jaws of the tool and position them so that the spine of the tool and the flanges are centered over the score line. Squeeze the handles together slowly and firmly and the piece should break along the score line. If it's not breaking on the score line, try spraying the wheel with a little aerosol lubricant. This tool takes some practice but once you've mastered it, you'll love it.

Contractors use the same method of cutting, but with a tool that sets on a table surface. A single handle is guided on two tracks to make a straight score line. Once the handle is lifted, two flanges drop down over the centered score line. Pushing the handle down exerts the pressure needed to cut the tile.

PHOTO 11

Virginia Bullman, *Nut Brown Maiden,* 1993. 28 x 16 inches (71.1 x 40.6 cm). Flower pot, ceramic thimble, concrete, broken pottery, pebbles; cast, mosaic. Photo by Cathy Seith Collection of artist

TILE SAW Tile saws use diamond blades for cutting ceramic, stone, and other vitreous materials. They're generally operated with water circulating over the blade to keep the blade cool while lubricating the cutting area and for washing away debris. It's not a tool you need to get started with mosaic, but once you've used one, it's easy to get addicted to the possibilities. They can be rented from your home improvement center; however, you can buy a good starter unit without making a huge investment.

DIAMOND SANDING BLOCK Whether you're breaking or sawing your tile, you may want to smooth sharp or chipped edges (photo 12). A diamond knife-sharpening stone or block actually accomplishes this quite nicely. Apply a little water to the surface of the diamond block. Hold the piece of tile so that the edge you want to smooth is at a 45° angle to the surface (photo 13). Press the tile firmly against the stone and move the piece in a circular motion four to six times or until the chips are removed and the edge is smooth (see photo 14). You only need to be concerned about the glaze surface. If you are grinding away and notice a fair amount of clay slurry on your block, you're working too hard and need to tip your mosaic piece more towards the glaze surface.

Sharp edges are of a particular concern if you're making a mosaic piece that will be out in the public. They need to be safe so people can interact with them. A mosaic buddy who has also done stained glass recently, suggested her glass grinder as a quicker solution for all those sharp edges. Most stained-glass grinders have a diamond-abrasive spindle but this one also had a diamond-coated disc. Water is also used in the process to cool, lubricate, and remove debris. Keep the 45° angle as you move the glaze edge across the spinning disc. It works beautifully, and saves a lot of time with tile preparation.

PHOTO 12

PHOTO 13

PHOTO 14

Adhering the Mosaic to Your Base Form

The selection of adhesive for any project depends on the surfaces you're trying to adhere and where and how the piece is going to be used. For concrete base forms that will be covered in ceramic and glass mosaic, I recommend a polymer-fortified thin-set adhesive. This is the same thin-set used for most tile-floor and swimming-pool installations. It's available in gray or white because it's a Portland-based product. If you're using glass gems or stained glass, the white adhesive will provide a uniform background visible through the transparent mosaic materials, and the colors will appear clearer and brighter.

The polymers added to thin-set are either in a powdered or liquid form. Some thin-sets have powdered polymers that are combined with the other dry ingredients during manufacturing. All you do is add water and stir to activate the polymers. Other companies market a two-part product that has a dry component and a white liquid containing the polymers, referred to by those in the business as milk. The two parts are mixed together without any additional water. It's very important to read the directions on the package of the thin-set you are considering. If a product needs to have separate liquid polymers added to it, and you only add water, the pieces may stick initially, but may not provide a waterproof bond to make them permanent.

Virginia Bullman, *Wife of the Farm,* 2002-2003. 60 x 36 x 36 inches (152.4 x 91.4 x 91.4 cm). Hypertufa over welded rebar armature and wire mesh, broken dishes; mosaic. Armature fabricated by Riley Foster. Photo by Andrew Ross

Materials and Tools

- Thin-set
- Container with lid for mixing
- Knife
- Ceramic clean-up tool (optional)

Instructions

1 To mix a thin-set that has the polymers incorporated in the powdered mix, put about ½ cup (112 g) of dry mix into a plastic container that has a tight-fitting lid and add 1 or 2 ounces (30 or 60 ml) of water. Stir the two ingredients together until the mixture is the consistency of peanut butter (photo 15). Allow the mixture to sit for about five minutes, and stir again to return to a creamy consistency. When not using the mixture, cover the container with the lid to extend the *pot life*, or the amount of time that the mixture is workable. If you are careful about covering the container when not in use, the mixture should be good to use for two to three hours.

PHOTO 15

PHOTO 16

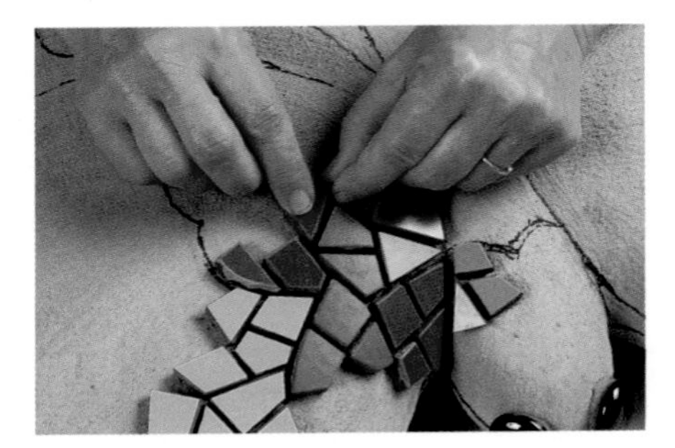

PHOTO 17

PHOTO 18

2 The technique I use to apply the adhesive to the surface is called *buttering*. I like to compare it to putting peanut butter on crackers (photo 16). I use old table knives to apply the mosaic adhesive. It's important to cover the entire surface of the mosaic piece, regardless of its size. You don't want to *starve* the joint by not applying enough adhesive. You also don't want to have so much adhesive on the back that it comes oozing around the piece as you stick it on. I like to generously cover the back surface of the mosaic material with adhesive and then lightly scrape off the excess on the side with the knife blade angled, so the remaining adhesive forms a low pyramid in the center of the piece.

3 When the mosaic piece is applied to the base form with firm pressure and a slight twist, the adhesive levels and a good bond is created (photo 17).

4 Any adhesive that oozes out is removed using the knife or a ceramic clean-up tool (photo 18). Allow the thin-set to dry for at least 24 hours before grouting.

Grouting

Grout is the cement-based material that fills the spaces between the mosaic pieces. In fact, your grout lines should be considered an important aspect of your mosaic design. The wider the spaces between your mosaic materials, the more dominant the grout color will be in your finished piece. You can also vary the width of your grout lines to help define shapes or add detail, such as using the grout line to function as the veins in leaves.

There are two different types of grout: sanded and unsanded. Typically, unsanded grout is used for interior applications where the spaces between tiles are uniform and the spacing between them is ⅛ inch (3 mm) or less. The consistency of unsanded grout is similar to plaster. When spaces are larger than ⅛ inch and irregular, you should use sanded grout. Most sanded grouts are mixed with liquid polymers or contain powdered polymers in the dry mix. The addition of polymers always adds flexibility to the finished material, making it desirable for exterior use.

Most home improvement centers will sell a brand of fortified sanded grout in a variety of colors. You can also use the Basic Sand Mix 3 on page 28 as a grout and add color to customize it. If you're grouting a large piece, be sure to measure your mix and pigments accurately to avoid ending up with a different shade each batch. Choose your grout color carefully. The color can set off a design or wash it out. Bright or dark colors of grout tend to fade when used outside.

Grouting a Mosaic Surface

The quality of your grouting can make or break an extremely time-intensive mosaic project. It's not that difficult to do, but it can be a little intense depending on the scale of your work and the variety of textures in your mosaic materials. The main thing to remember is that it's a process that must be completed in one continuous work session. You can't stop and go out for dinner in the middle of grouting a project. It's best to get organized and block out the time needed to do a good job. If you can physically hold the piece you want to grout, such as a small sculpture or planter, allow about three to four hours for the complete grouting process from setup to cleanup. If the piece is larger, enlist the help of assistants, and plan on a four- to six-hour work session.

K.C. Linn, *Foolish Thinking,* 2004. 37 x 18 x 18 inches (94 x 45.7 x 45.7 cm). Vermiculite, Portland cement, and polyfibers over fiberglass-covered polystyrene foam, cement bond; sponge stamped. Photo by Elder G. Jones

Materials and Tools

- Container to mix grout
- Part measuring container
- Grout mix
- Rubber gloves
- 4 ml plastic
- Sponge
- Water container
- Clean terry cloth rags
- Spray water bottle
- Blanket (optional)

Instructions

1 Wearing rubber gloves, mix the grout following package directions or the custom recipe. If you have a large bag of grout but only want to mix a small amount, the basic proportions are 4 to 1—four parts of powdered mix to one part water. When mixed properly, grout should have the consistency of sugar-cookie dough—a no-slump consistency that is creamy and doesn't crumble. Allow the mixture to slake, or sit about five minutes so the moisture can be absorbed by the grout particles. Mix again so it's workable and even in texture.

2 Take a small handful, and using the heel of your palm, compress it into the surface of your mosaic using a small circular motion. Continue to rub the mixture until most of the material is off of the mosaic surface. Take your next handful, and overlapping the initial area slightly, apply it in the same manner. Continue to do so until the entire mosaic has been filled with grout.

3 Allow the grout to set up until firm. For a smaller piece, you may need to wait 20 minutes or so. For larger pieces, it may have taken so long to apply the grout that it will be firm enough to proceed with the next step in the area where you started.

4 Level the grout between the mosaic pieces. Dip the sponge in water, squeeze it almost dry, and rub firmly, in a circular motion, over the mosaic surface to further compact the grout line and to remove deposits of grout from the mosaic. If you rub too hard in the direction of the spaces, you'll begin to remove the grout. Rubbing in a circular motion will help avoid this. Continually rinse and squeeze your sponge. Rotate the sponge as you work, always turning it to a clean side. Don't be timid about

rubbing quite firmly. If you notice low spots or even holes in the grout lines, add more material, let it set up, and then sponge again as recommended. After you have leveled the grout and removed the deposits of excess material from the mosaic pieces, stop sponging.

5 Allow a haze to form over your mosaic material as the grout residue dries on the surface. The more vitreous your mosaic material is, the longer the process will take.

6 Once the haze has formed, use a clean terry cloth rag to buff the mosaic surface in a circular motion. This is the rewarding part. Throughout the process so far, the piece has looked dirty; now it begins to sparkle. To thoroughly clean the mosaic, you may need to wrap the terry cloth over your finger and clean individual pieces. Reposition your terry cloth as you're working so you're always using a clean surface. Change to a clean terry cloth rag as needed. Three-dimensional pieces may need to be rotated in order to properly finish the grouting process. Do this with care, as the grout is still very fragile. You might want to create a soft work area by covering a folded blanket with plastic.

7 When you're finished, follow the recommended curing procedure listed on the grout package instructions. Remember, this is a cement product and it must cure properly to ensure strong grout joints. Two things that will cause cracking in grout are drying out too fast, and not compressing the material firmly into the spaces during the initial application.

Reverse-Cast Mosaic

This process is an example of an indirect mosaic technique. The mosaic is created on one surface and transferred before becoming affixed to another surface. By starting with the design face up on your work surface, you have the flexibility of arranging and changing your mosaic pieces. You can even use materials of differing heights and still achieve a flat finished surface.

Materials and Tools

- Concrete Mix: Premixed Sand/Topping Mix, Mortar Mix, Commercial-Grade Mason Mix, or Mix 3
- Mold
- Paper for template
- Pencil
- Permanent marker
- Assorted frost-proof materials
- Tile nippers
- Tile cutters
- Hammer
- Newspaper
- Diamond sanding block (optional)
- Scissors
- Clear adhesive shelf paper
- 2 work boards of the same size
- Mold release agent
- Hardware cloth
- Aviator shears
- Container to mix concrete
- Screed
- 2 ml plastic
- Sponge
- Container for water
- Soft nylon brush or nylon pot scrubber

Instructions

1 On a piece of paper, trace or measure the inside top surface of the mold you'll be using to cast in. Draw another line ¼ inch (6 mm) inside that line. This space is needed to soften the edges of the mold line to finish your casting. The inside area is the template you have to create your mosaic design.

2 Organize a variety of colors, shapes, and materials inside the template area, or draw a more specific design.

NOTE When you're drawing a design for mosaic, develop it in shapes rather than lines. Sketch out your design on the paper template. Use a sharp permanent marker to simplify or clarify your sketched lines. Select, break, cut, or nip your mosaic pieces, and begin to arrange the pieces for your design as if you were creating a jigsaw puzzle. Lay the pieces on top of the design template with the glaze side up. As long as the mosaic materials have a flat surface, you need not be concerned about the variation of thickness. Once the mosaic is transferred and laid upside down, the top surfaces will be level.

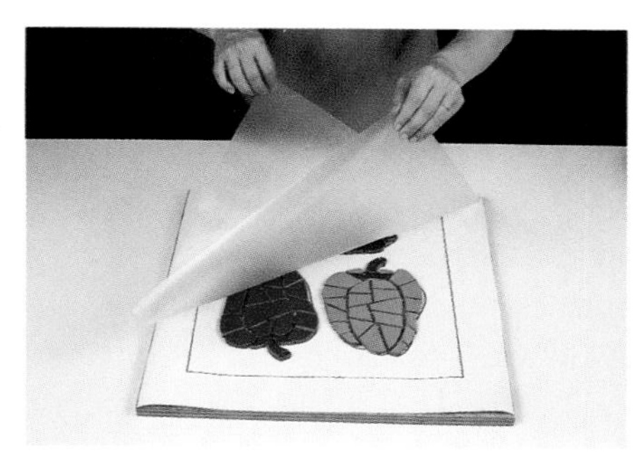

PHOTO 19

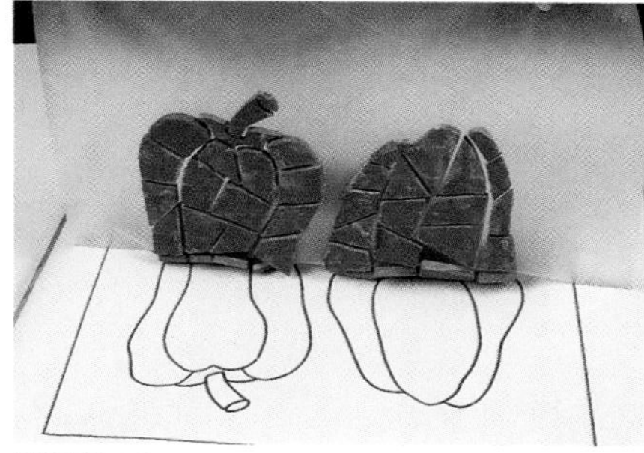

PHOTO 20

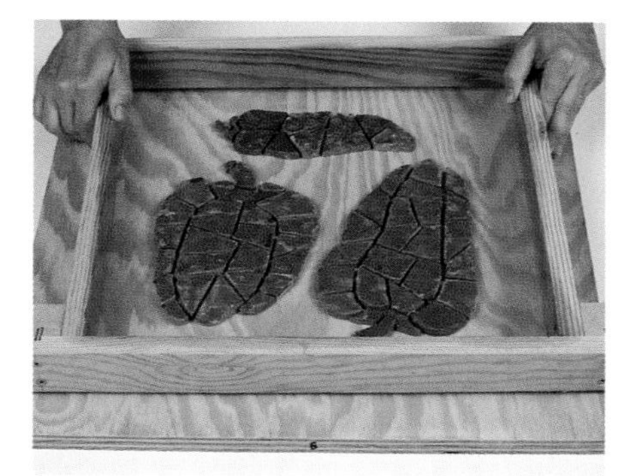

PHOTO 21

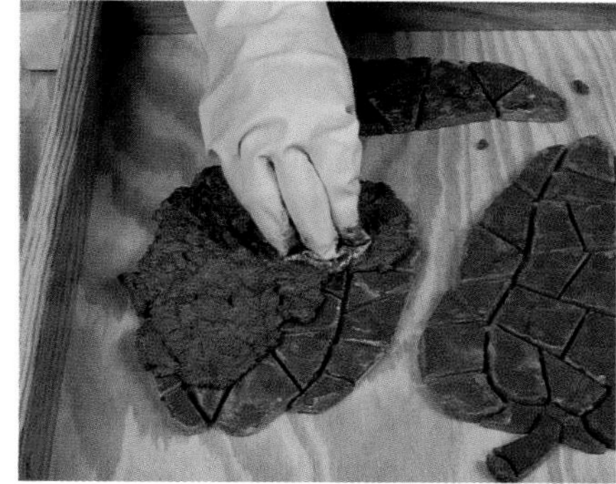

PHOTO 22

ANOTHER TIP When using glass gems, place the flat side up. You'll have a nice round glass dot in your finished casting. If you use the rounded side up, concrete will come up over the edges, leaving you with just a speck of color. As you organize your mosaic material, make sure that you leave at least a ⅛-inch (3 mm) space between each piece. You can fill the template with consistently spaced tiles, or you can simply position a few pieces of mosaic materials. The cast concrete will create the background color.

3 When you've finished positioning your design, cut a piece of clear adhesive shelf paper to about the size of your work board.

4 It helps if you have a second set of hands to help you with the next two steps, but you can do it solo if you need to. Remove the backing of the shelf paper, and hold it with the sticky side down. Hold it so it creates a "U" shape (photo 19).

5 Let the bottom curve of the shelf paper touch the center section of the mosaic design, and gently ease the sides over the rest of the design. If you try to hold the adhesive paper flat to apply the design, the static electricity, as well as the awkwardness of the position, will cause the design pieces to shift.

6 Rub or *burnish* the adhesive shelf paper over the surfaces of the mosaic materials without adhering it to the paper template.

7 After you're sure that the pieces are attached, pick up the shelf paper, which is now holding your mosaic pieces (photo 20) and turn the design upside down onto one of the work boards. By cutting the clear adhesive shelf paper the same size as the work board, you have provided a plastic surface and the casting won't stick to the board.

8 Apply the mold release to your mold, and center your mold over the design (photo 21). If you keep the mold release off the edge of the mold, the adhesive shelf paper will also help to keep the mold in place during the casting process.

9 With the aviator shears, cut a piece of hardware cloth ½ inch (1.3 cm) smaller than the perimeter of the mold, and set it aside.

10 Mix the concrete. Take a small handful of it, and press or tap firmly, but carefully between the mosaic pieces (photo 22). You don't want the pieces to come loose from the shelf paper, and you don't want the mold to move. Continue to do this until the spaces between the mosaic materials are filled and the bottom surface is covered. Make sure that you have tapped the concrete firmly into the corners.

11 Fill the mold halfway. Position the hardware cloth so it doesn't touch the sides of the mold, and continue filling.

12 Follow steps 9 through 13 under Casting a Simple Mold or Form (see pages 38 through 39).

PHOTO 23

PHOTO 24

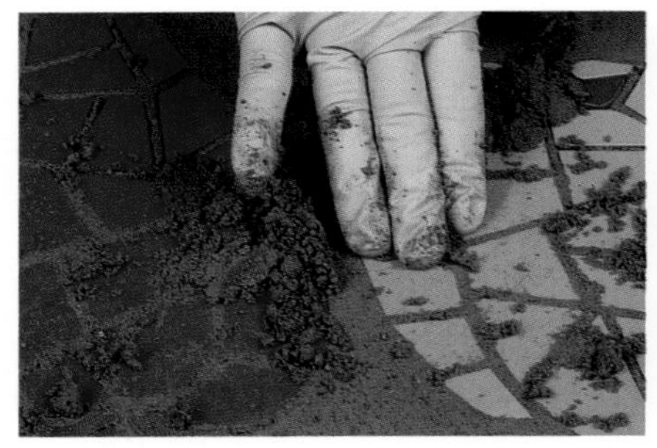

PHOTO 25

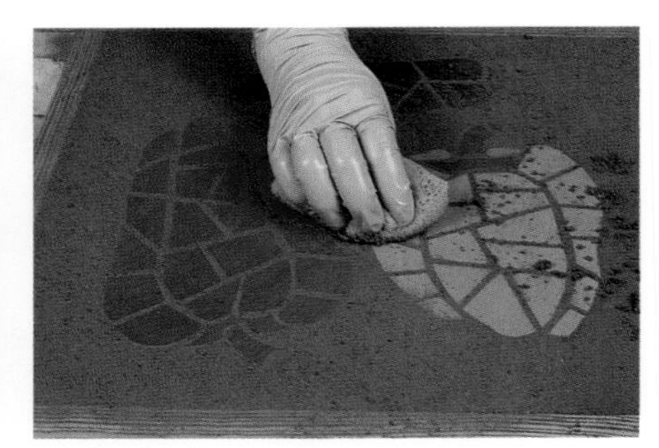

PHOTO 26

PHOTO 27

13 Place a piece of plastic, then the second work board on top of the casting, so that it's now sandwiched between the two boards (photo 23). Depending on the size of the casting and your strength, you might want an assistant to help you turn it over. Hold the two pieces of wood together firmly and turn it over (as if you were taking a cake out of a baking pan).

14 Remove the shelf paper (photo 24). If there are voids in the interstices, take some of the reserved concrete and rub it into the surface in a circular motion (photo 25).

15 With a slightly damp sponge, continue to work over the surface in a circular motion (photo 26).

16 After the spaces have been filled, rinse and ring out the sponge, and start to clean the design area, continually turning the sponge to clean surfaces as you wipe across the mosaic. Rinse the sponge frequently as you do this. Use the sponge on the plain concrete areas to even out the surface texture.

17 Clean off the edges of your mold and then remove the casting from the mold (photos 27).

18 Carefully smooth the sides, edges, and mold lines with the damp sponge (photo 28)—don't worry if there's a haze on the mosaic surface.

19 Cover with plastic and leave undisturbed overnight.

20 After 24 hours, use the soft nylon brush or nylon pot scrubber to clean the mosaic surface (photo 29). Carefully turn the casting to file the bottom edges, remembering that the concrete hasn't reached its full strength yet (photo 30).

21 Keep the casting wet for five days to cure (photo 31).

I've used a steppingstone in this example, but this reverse-cast method can also be used to create decorative panels or wall hangings. Consider making a thinner mold and using extra reinforcing material to minimize the weight. You might also want to include hanging loops in the backs of these pieces when you cast them to make installation easier (see page 49).

PHOTO 28

PHOTO 29

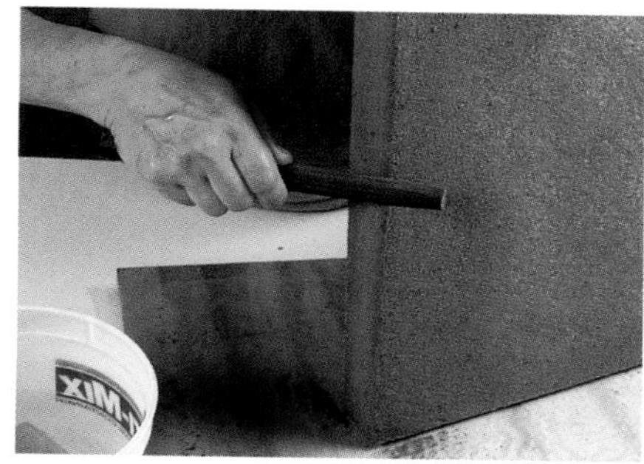

PHOTO 30

PHOTO 31

Virginia Bullman and LaNelle Davis, *Wilson Train Lady,* 2004. 60 x 60 x 27 inches (152.4 x 152.4 x 68.6 m). Hypertufa mix concrete over heavy rebar armature, pottery shard mosaic. Photo by Andrew Ross Location: Amtrak station in Wilson, NC

Embedding

Embedding is the process of taking an object and pressing it into wet concrete so that it becomes a part of the surface. Depending on the personality of the maker, embedded pieces can either look somewhat controlled or extremely spontaneous. Amazing self-taught artists have created magical environments worldwide using the embedding process by embellishing their sculptures and structures with everything from beer bottles to Sacred Heart medals and from china to plumbing fixtures (see pages 115 through 117).

Marvin and Lilli Ann Killen Rosenberg, *Sidewalk Medallion.* 3 feet (0.9 m) in diameter. Concrete, high-fired stoneware inserts, set in exposed aggregate sidewalk; mosaic.
Photo by artist Location: Ashland, OR

Embedding on a Base Form

Part of the attraction of embedding is that it can be combined with several different forming techniques. It can be incorporated into the final application of concrete to base forms, such as the bowl and pedestal base of a birdbath, or utilized as a decorative method in castings, such as steppingstones or sidewalks (see the photo at left).

Materials and Tools

- Concrete Mix: Premixed Sand, Topping Mix, Commercial-Grade Mason Mix or mix 3
- Slurry Mix: Mix 20 or Mix 21
- Mosaic materials and tools (pages 95 through 98)
- Container to mix slurry
- Paintbrush
- Container to mix concrete
- 6-inch (15.2 cm) piece of ⅝-inch (1.6 cm) diameter wooden dowel
- Sponge
- Nylon brush
- Small stainless-steel brush
- Wooden craft stick
- File (optional)

PHOTO 32

Instructions

1 Make your base form.

2 Prepare your materials for embedding. You may want to design a recognizable image or just have a selection of materials that you'll use randomly. It's much easier if you prepare more materials than you'll need rather than having to stop and prepare more when you're almost at the end of the project.

NOTE Often I'll decide on a few different colors of tiles and a few additional elements to provide texture. I'll prepare the tiles by cutting or breaking them into certain shapes or sizes so that I have a good supply of materials on hand and organize my materials in low-cut cardboard boxes or trays. Once the concrete is applied, I go to work in a method I like to call "planned randomness." I plan on a limited palette or selection of decorative materials, and apply them in a random way. The limitation on the variety of elements helps to give the piece a sense of unity, and often a rhythm or pattern is created subconsciously.

3 Mix the slurry and concrete. Paint the slurry onto an area slightly larger than where you plan to apply the concrete (photo 32).

4 Apply an additional layer of concrete using the patty technique described on page 49 and 50 (photo 33). The layer of concrete only needs to be as thick as the thickest material you're embedding. This works easily on the inside of a bowl, which is basically a horizontal surface, but gets trickier when applied to a vertical base. When you apply this layer to a vertical surface, try to avoid creating air pockets. Easing the cement onto the form by pushing firmly rather than trying to slap it on flat will help push out any trapped air.

5 Apply as much slurry and concrete as you think you can embed within 20 minutes or so. Periodically apply more slurry and concrete so the concrete has a chance to set slightly before you begin to embed. Don't embed to the edges of the applied concrete. You need to leave at least 1 inch (2.5cm) to give you room to overlap your next application.

PHOTO 33

Embedding Tips

Due to displacement, you may see a small space between the embedded object and the concrete. You can either add an additional pinch of concrete, or take a wet sponge, squeeze out most of the water, and sponge over the concrete toward the center of the embedded object. This should move the concrete into that small space. If you're embedding a large, porous object, first dip it in water and then blot off the excess moisture. Adding moisture to the porous object before trying to embed it will keep it from absorbing too much water from the concrete.

When using hollow materials like rounded glass pieces or seashells, fill any open spaces with concrete before embedding (see photo below). This will help to reinforce the piece. If the object gets broken and is not filled, it creates a space for the water to collect, and during a freeze/thaw cycle, more damage could occur. If the object gets broken, and it had been filled with concrete, you'll have a concrete fossil on the surface of your piece that will still look interesting.

If you find that your concrete is beginning to set up, and things are not tapping in so easily, cut out a small shape from the cement on the base form approximately the size of the decorative element, add a dab of slurry, and then tap in the piece. Additional pinches of concrete can fill any remaining spaces.

If embedding a bowl, plan a work session long enough to complete the design to the edge of the bowl. If you stop work in the middle of the bowl and return to it the next day, there may be a visible line or ridge distinguishing the two applications. If you want to embellish both sides of the bowl, do one side the first day, allow that to set up for 24 hours, and do the other side on the second day.

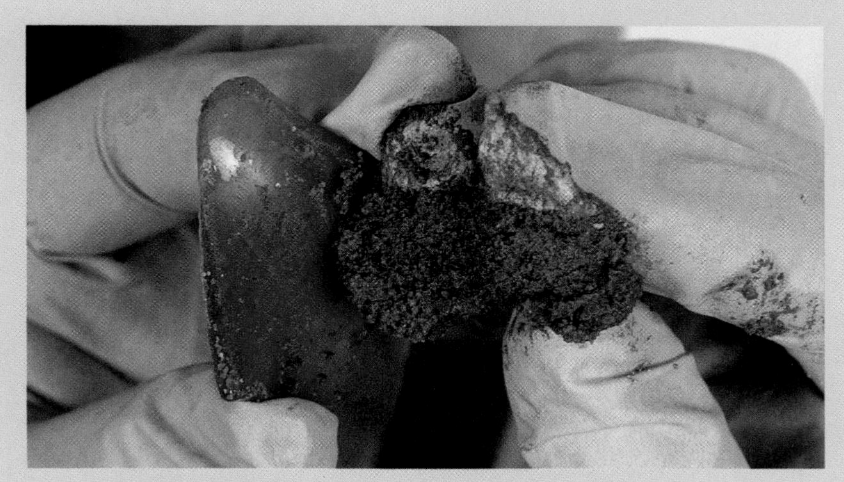

Filling a glass shape with a pinch of concrete

PHOTO 34

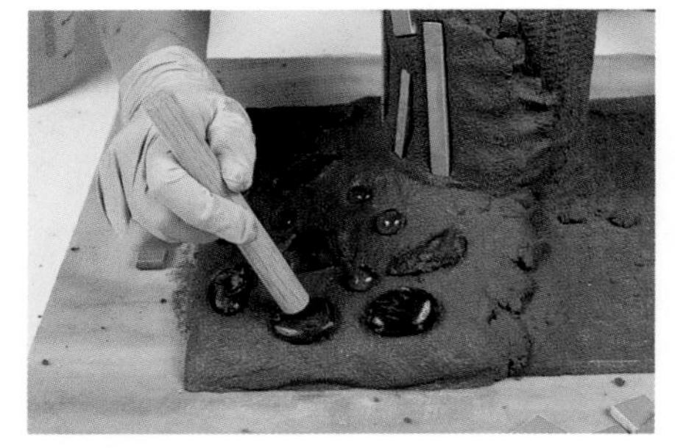

PHOTO 35

PHOTO 36

6 Arrange your decorative pieces on the concrete surface (photo 34). When you're satisfied with the placement, use the piece of wooden dowel to firmly tap the pieces into the wet concrete (photo 35). Embed the pieces so the surface is level with the concrete.

7 Use the damp sponge to clean off deposits of concrete and smooth the surface (photo 36). Remember to rinse your sponge frequently and rotate it to clean surfaces as you work. Don't be concerned if a thin layer or film of concrete remains on the surface. Continued sponging is only going to smear more concrete over the surface, not clean it completely, so stop sponging.

8 When you're finished working, cover your piece with plastic and leave it undisturbed overnight. The next day, take the piece outside and spray it with water. Use the nylon brush to clean the film off the decorative pieces. You might want to use a toothbrush-sized stainless-steel brush to clean more difficult or textural areas, but be careful that you don't scrub so hard as to scratch either the material or the concrete; it feels solid, but it hasn't yet reached its full hardness. Chunks of concrete can be pried off with the wooden craft stick or carefully removed with a file. When you've finished cleaning, rinse the piece with water one last time.

9 If you're able to clean your piece within 12 to 24 hours of embedding, you'll find that most of the cement residue can be removed with water and elbow grease. If you're unable to work on your piece until some time later, it may be necessary to use a chemical for the final cleaning (see page 26). Cure for at least five days.

Coloring Concrete

The two main ways of coloring your concrete are applying a colorant to the surface, and adding a colorant to the mix to integrate it throughout the concrete. The technology and resulting effects for coloring concrete have expanded significantly over recent years. Experimentation will be your best resource in developing creative combinations for personal expression. In the same way that I suggested you customize your mixes and keep track of the results in a journal, record the effects of your experimentations in color as well.

Applied Color

The two basic choices with applied color are paints or stains. In most cases, coloring processes applied to the surface

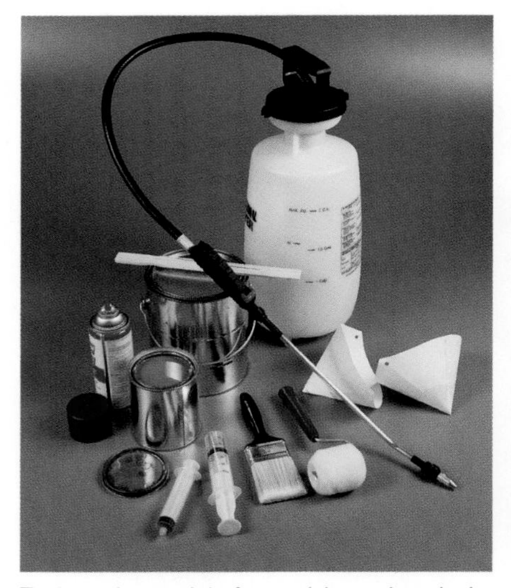

Tools and materials for applying color, clockwise from top right: garden sprayer with metal tip, disposable strainers, small paint roller, paintbrushes, veterinary horse syringe, marinade syringe, latex-based masonry paint, spray concrete sealer, sealer, stirring stick

A painted concrete surface

of concrete should only occur after the concrete has thoroughly cured and is dry, to avoid efflorescence, blistering, or peeling. In some cases, you may even want to chemically clean your piece so that any of the efflorescent salts that have emerged to the surface are removed before applying your color. Follow the surface preparation instructions provided by various manufacturers.

PAINTS

Acrylic masonry paints can be used on concrete to create a solid opaque finish. A good masonry primer will help to prepare the surface by providing a good base and bonding coat for your final finish. Paints also flatten the surface, filling in the pores and texture of the concrete (see photo above).

Artists use the pure color pigments used to mix house paints in two ways. In some cases they add the paint colorants directly to the concrete mix to produce an inherent color as described below. Kathy Hopwood uses them to mix her own paints by combining paint pigment, a concrete bonding agent, and water (see photo above right).

Kathy Hopwood, *Autumn's Delight,* 2003. 12 x 14 x ½ inches (30.5 x 35.6 x 1.3 cm). Cement, latex pigment; cast. Photo by Andrew Ross

Paint as Stain, or Color that Rocks

It's not always necessary to use paints at the same consistency as they come out of the can. While experimenting with painting the faux boulder (see page 132), we followed several suggested methods and came up with this combination that worked for us. Start by selecting a palette of colors for your rock. We used a neutral light gray, a darker shade in the same color family, rust, and black.

Materials and Tools

- Concrete boulder
- Pressure washer
- 4 or more colors of masonry latex paint
- Mixing containers
- Mixing sticks
- Disposable paint strainers
- Garden sprayer with metal spray tip
- Spray water bottle
- Air sprayer (either from a compressor or canned air)
- Large plastic syringe (a marinade injector or the type that is used for horse vaccinations)

Instructions

1 Use a pressure washer to wash any loose debris from the surface of your boulder.

2 Thin the lightest color of paint to about the consistency of milk, then strain it through a commercially available paint strainer. Put the paint into the garden sprayer and spray the boulder completely, taking care to spray in all the nooks and crannies so that all the concrete is covered. Give it a second coat if needed.

3 Clean out the sprayer and then add the thinned and strained darker shade. This time just spray the

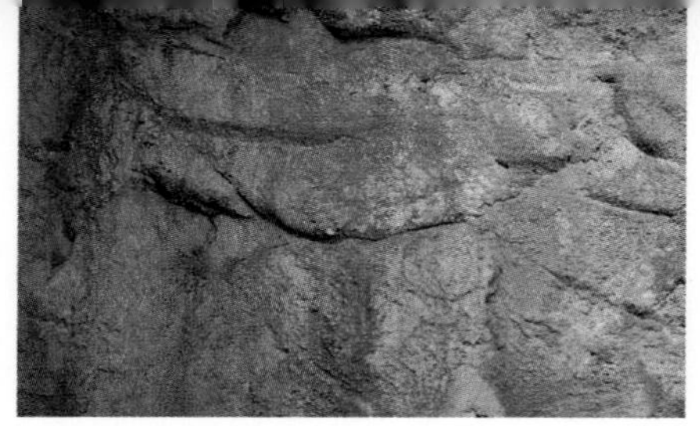

PHOTO 37

PHOTO 38

areas you want darker. These coats should be worked wet-on-wet so the colors blend.

4 The rust color can be applied either with the sprayer or with the syringe, depending on the effect you are looking for. By now, some of the paint may be drying. As you apply the rust paint you want to avoid any telltale drips; feather the application by misting water onto the area with your spray bottle (photo 37). If you want the paint to travel in a specific direction, use your air spray to force it into a specific ridge or down a surface.

5 Using watered-down black paint adds to the rock illusion by helping to "deepen" the crevices (photo 38). Fill the syringe with the paint. Direct the tip into a crevice and pull it along as you are dispensing the paint. The black streaks can be lightly feathered with your water bottle or air source.

6 Enhance color areas as needed.

STAINS

Several varieties of stains work well on concrete. Some companies produce opaque acrylic stains. They are commercially available in a wide range of colors and provide solid coverage but allow more of the surface texture to come though.

Transparent concrete stains are another option. These stains show, even highlight, the concrete texture (see photo below). Some concrete artists suggest using a water-base or an opaque acrylic stain originally formulated for wood, but there is no long-term guarantee of their lasting effects. Apply the stain generously using a stiff brush, and then wipe off the excess with a damp cloth. Multiple coats can be applied to add color density or variation.

Acid stains, however, are specifically designed for concrete. As their name implies, there is an acid component in the stain that etches the concrete, allowing the stain to penetrate the concrete surface. Johan Hagaman (see photos on pages 51, 56, and 73) uses acid stains in combination with other coloring techniques to arrive at her rich surface. Read the manufacturer's instructions, and be careful to follow all safety precautions.

Transparent concrete stains can show texture.

Kem Alexander's concrete eggs

SEALING

Whichever color application you use, consider sealing the finished surface of your concrete to reduce water penetration and staining. Most readily available sealers are acrylic or water-based, but these stay on the surface, and if applied too heavily they will give an artificial glossy appearance. They may even yellow over time. Penetrating sealers are recommended. Kem Alexander's eggs, see photo above, have been finished with a penetrating sealer. They almost seem to glow from within.

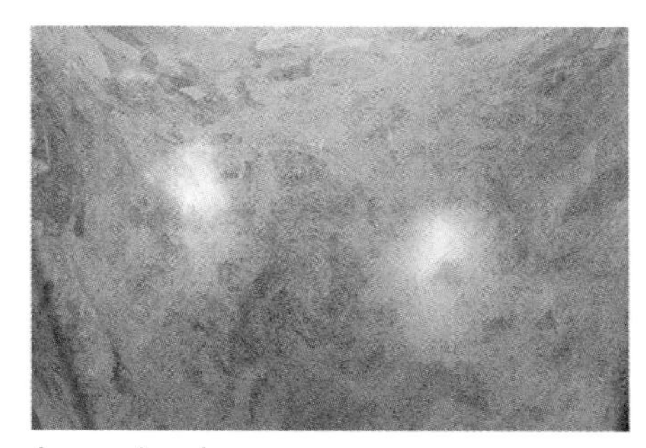

A waxed surface

WAXING

Waxing provides a final coat that will inhibit staining; however, you will end up with a glossy surface. The most recommended wax to use is a paste wax that includes carnauba. When your concrete piece is clean and dry, apply the wax with a clean cloth. Rub it into the surface, then buff with a soft clean cloth (see photo above).

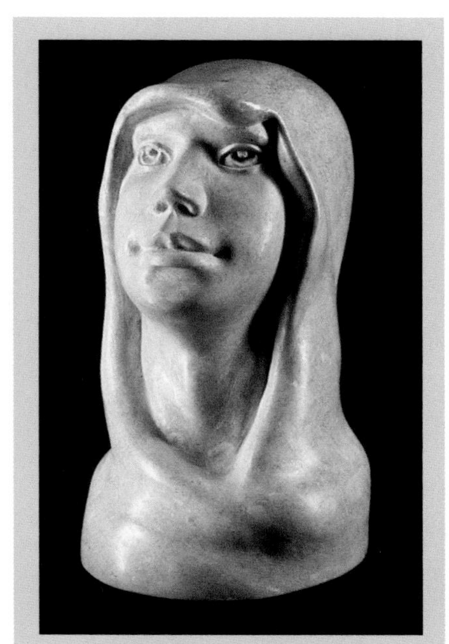

Lynn Olson, *Lori,* 2000. 10 ½ x 5 ½ x 7 inches (26.7 x 14 x 17.8 cm). White Portland cement, steel wool fibers, steel rods; direct modeling construction, surface filed and sanded, sealed with methyl methacrylate. Photo by artist

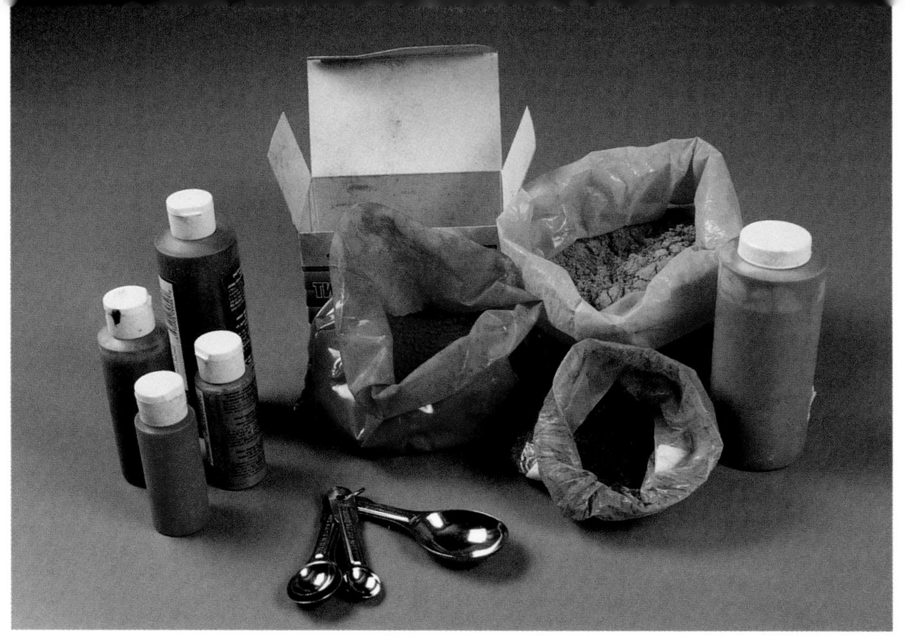

Colorant materials, clockwise from top: mason stain pigments, liquid concrete color, powdered concrete color, measuring spoons, acrylic paints and pigments

Inherent Color

Adding color to your concrete mix gives you an inherent color—a color throughout. There are limitations when adding color to concrete. As mentioned in earlier sections of this book, Portland cement is caustic. Some pigments that work well in paint are attacked by the alkali in cement, resulting in a breakdown of the color over time. Also, colorfastness is a consideration. You need to know how well your color agent can endure exposure to the sun and other natural elements.

PIGMENTS AND COLORANTS

It's not too difficult to find concrete color additives in earthy tones: rust, ochre, brown, charcoal, and sometimes green. These are easily found at your home improvement center in either a powdered or a thick liquid form. Other colors like red, orange, and various shades of blue are derived from chemical compounds that are more expensive. My best source for locating these pigments has been through a ceramic supply company that sells powdered mason stains. Of course, there are concrete suppliers who also sell powdered pigments—the question there may be the quantity in which they're willing to sell them.

Getting a deep, rich color may also present a problem. The maximum amount of pigment that should be added is 10 percent of the weight of the cement in the mix. A higher percentage of pigment can weaken the mix, since more water is needed to wet the pigment particles, which increases the water/cement ratio of the mix. To avoid this, you might want to create a base form without pigment and then apply a topcoat with pigment. This creates a stronger structure with a deeply colored surface. This is essentially what I did to create the bowling-pin form for the totem on page 168. The sealer that was used on the finished cured form also helped to deepen the color.

PHOTO 39

PHOTO 40

PHOTO 41

PHOTO 42

Pigments added to white Portland cement mixes will result in cleaner looking colors. Compare the two color samples shown in photos 39 and 40. Both batches of concrete were mixed with exactly the same amount of powdered pigment. In photo 39, white Portland cement was used, and in photo 40, gray. In both cases, the powdered pigment was mixed thoroughly before adding water. To achieve a variegated color, mix the concrete first and then add the pigment, mixing just enough to achieve a marbleized effect.

Another technique that can be used on a horizontal surface for achieving color mixes or variations is to shake or dust the powdered pigment over the surface of fresh cement. In photo 41, I cast, screeded, and troweled the surface, and then waited for the bleed water to evaporate before shaking on black and red powdered pigments. I then troweled the surface again until the pigment was worked into the cement and I had achieved the desired blend. I did have to pause when bleed water resurfaced, and I wiped my trowel clean several times during the process to avoid getting the colors too muddy.

AGGREGATES

Once you become familiar with the range of aggregates available, it's easy to understand how they can affect the color of your concrete—from seeding on the surface to being integrated with your mix like the bench project on page 137. In most cases, the addition of aggregate plays a more significant role in the color scheme if you grind or polish your concrete (photo 42).

Andrew Goss, *Red-Line Bracelet,* 2004. 3 x 3 1/2 x 1 inches (7.6 x 8.9 x 2.5 cm). Ferro-cement (mortar around steel mesh armature), dye; inscribed. Photo by artist

Inlaid Color

The beauty of inlaid color combines contrasting pigmented cement with elbow grease. Andrew Goss uses this technique frequently. To achieve a surface similar to that in photo 43, follow these steps. It's best to work this process while the concrete is fresh, finishing each application within 24 hours of the last, so that the bonding is more complete.

PHOTO 43

Instructions

1 Start with a base form that is refined and almost at its finished state (see photo 44).

2 Add a white layer, cover, and let it set until firm, about 24 hours. Smooth it, and then score and texture it with coarse files, knives, or gouges. Brush away loose particles, rinse, and wipe off excess water.

3 Add a gray smooth coating (Mix 19), rubbing it into the textured areas. Let that set, covered for 24 hours. Then wet sand through the gray layer to reveal the white with the dark inlaid lines. Start your sanding with 180- or 220-grit wet sandpaper and work up to finer grits of 320 or 400.

4 Cover the piece for 5 days before wet sanding with 600-grit paper, then rinse and dry.

5 Paint the piece with penetrating sealer, let dry, and then wax.

PHOTO 44

Remember to keep your work covered between work sessions and to keep the surface damp by misting with water. Of course, your contrasting layers can be any color you choose, but the final inlaid contrast will only occur by working patiently to achieve the desired effect.

A similar process is backfilling, but this usually refers to voids that are left in castings rather than marks that are made on purpose. Backfilling is covered on page 94 as part of the polishing process.

Colored by Nature

I've already confessed to my love of the raw material, but sometimes nature steps in and adds a special touch to concrete. You can also lend her a hand.

Rust

The beauty of rust has reached new heights with artists like Kem Alexander, as shown in photo below. She has collected rusty hardware of all descriptions to incorporate into her concrete sculptures. All of the metal she uses is cleaned, dried, and free of any oily residue. She prepares her metal like a mosaic artist preparing tile. Everything is planned and cut to length before starting. The rust from the objects migrates into the body of the concrete. In some instances this may not be desirable, but in the hands of this artist, it truly is a thing of beauty.

Kem Alexander's concrete and rust

Moss

Mossy troughs are a sought-after accent in many gardens as a symbol of timeworn beauty. Okay, so you just finished your hypertufa trough two weeks ago and you'd like that timeworn look sooner rather than later. Let's feed the process.

Materials and Tools
Moss
Buttermilk
Scissors
Blender
Paintbrush
Hypertufa container

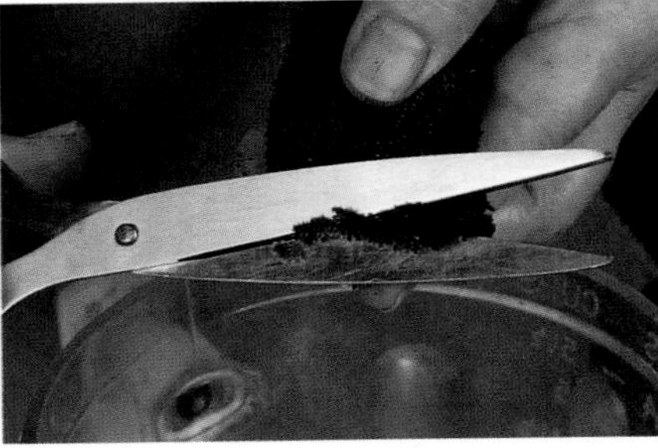

PHOTO 45

Instructions

1 Gather some moss. Cut a cup or so of moss with scissors into the blender so that dirt and pebbles aren't added to the mix (photo 45).

PHOTO 46

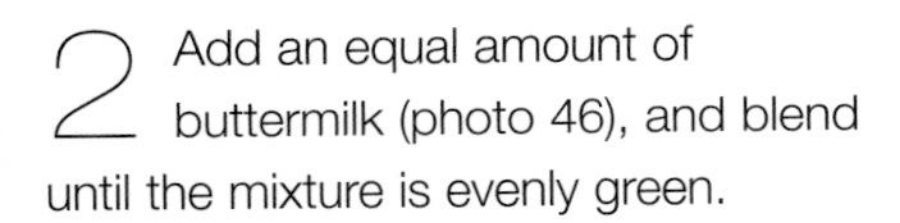

2 Add an equal amount of buttermilk (photo 46), and blend until the mixture is evenly green.

3 Paint the mixture onto your container (photo 47).

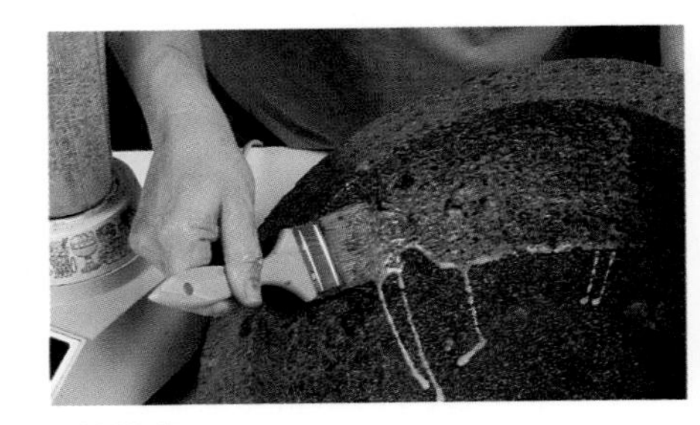

PHOTO 47

4 Cover with plastic for a few days and place in a cool location.

NOTE During my first attempt at growing moss, I moved my just-sprouting bowl, still wrapped in plastic, outside. The direct sun created a greenhouse effect, and I fried my moss. Thought you might want to learn from my mistake.

5 Unwrap and mist regularly. Just mist—watering will wash away the fragile starts (photo 48).

PHOTO 48

Make Your Dreams Concrete!

My inspiration for creating in concrete started decades ago during my freshman year in art school when one of my instructors showed us his slides from his travels to record the work of self-taught artists. These images had a profound effect on how I viewed art and the creative process.

Willem Volkersz has continued his travels, spanning the ocean in his search to capture the work of visionary artists. His images, used here, allow me to share with you the work of artists who have followed their hearts and visions to make their dreams concrete.

Left:
Ed Galloway (1880–1962)
Totem Pole Park, Built between 1937–1962
Foyil, Oklahoma, USA

Ed Galloway always thought of his building projects as a hobby that he started after his retirement in 1937. He used an armature system of stacked sandstone and scrap metal for his constructions, which he covered in concrete bas-relief and then painted. The main structure is shaped like a teepee and is 90 feet (27 m) tall, topped with a carved cedar pole. The base, shaped like a big turtle, measures 18 feet (5.4 m) in diameter. His imagery was fed by his love for *National Geographic* magazines and the postcards of totem poles his daughter would send him from her home in Alaska.

Right:
Bruno Weber (1931–)
Weinrebenpark (Grapevine Park)
1962–to present
Dietikon, Switzerland

It was Bruno Weber's reaction to the urban development of nearby Zurich that inspired him to create a contrasting magical world that would feed the imagination. He has defined three major gardens with a population of animals and mythical creatures. A tower house, pavilions, and water features provide added dimension to the dream-like setting. Together with his wife, Mariann Weber-Gordon, they have developed their own building technique using waste molds to cast pieces which are then inlaid with mosaic. Larger structures are planned with the help of engineers, but for the most part Weber's construction is intuitive.

Ferdinand Cheval (1836–1924)
Palais Ideal, 1879–1912
Hauterives, France

Ferdinand Cheval began collecting tufa rocks for his garden when he was a rural postman. Eventually, they ended up being embedded in concrete reinforced with wire. His inspiration for the Palais came from daydreams he had on his repetitious route. He combined grottos, towers, and sculpture into a four-sided structure that is 39 feet high x 86 feet x 46 feet (12 x 26 x 14 m). Cheval's Palais is sometimes referred to as "convulsive beauty" and is linked with the surrealist movement, even though his dream building was much more intuitive than philosophical.

Simon Rodia (1879–1965)
Watts Towers, mid 1920s–1954
Los Angeles, California, USA

Simon Rodia's work experiences as a quarryman, road builder, cement mason, and tile-factory worker are reflected in these wonderful towers. He covered his bent-metal and wire armatures in concrete, embellishing them with stamping and by embedding found materials of all descriptions. In 1959 the city threatened to tear the towers down, but they survived, and in recent years have gone through an extensive renovation.

Albert Gabriel (1904–)
Le Jardin Sculpte d' Albert Gabriel (the Sculpture Garden of Albert Gabriel or Chez Gabriel), 1967–to present
Brizambourg, France

Albert Gabriel always wanted to be a sculptor, but became a farmer. When he retired in 1967, he was able to follow his dream. Working with colored concrete over metal armatures, he's made more than 100 figures of celebrities, both real and fictional, as well as everyday people.

Herman Rusch (1885–1985)
Prairie Moon Sculpture Garden and Museum, 1952–1974
Cochrane, Wisconsin

Herman Rush purchased the Prairie Moon Dance Pavilion to transform it into a museum. The barren grounds led him to make a concrete and stone planter. Soon, this self-taught artist was creating gardens and building sculptures. Chiseled rocks and bricks went to build the conical supports of the arched fence. He embedded the concrete with bits of broken bottles, crockery, shells, and mirror, and added color by mixing it directly into the fresh concrete or by painting the surface. At age 94, he sold the property so he could spend more time fishing and fiddling.

Eddie Owens Martin, AKA St. EOM (1908–1986)
Land of Pasaquan,
approximately 1959–1982,
Buena Vista, GA

Eddie Martin would return home occasionally from New York City to help his mother with her small farm in rural Georgia. Once when seriously ill, he had a vision that he would be known as St. EOM, the first of the Pasaquanians. After his mother died, he returned to Georgia and began creating his compound. He defined the garden with varying heights of concrete-covered walls. His art was inspired by his connection to the cosmos and his research into Eastern religions, ancient art, and tribal cultures.

CHAPTER FOUR

The Projects

Now that you've learned the basics of working with concrete in Chapters 2 and 3, it's time to mix some serious concrete! The projects in this section are grouped according to skill level—Beginners, Intermediate, and Advanced. If you've never worked in concrete, you'll be more comfortable working through the first eight projects, which are on the Beginner level. The next eight Intermediate projects let you take your basic skills and expand them with techniques that may take a little practice. Finally, the Advanced projects, for those of you with some experience under your belt, challenge you to take what you've learned and test your skills.

Before you begin, flip through the projects to find one you feel comfortable making. Make sure to read over the Tools and Materials list and the instructions. Many of the projects refer back to the techniques described in Chapters 2 and 3. You'll want to take into consideration the tools and materials for those techniques in addition to those listed for each project. Each project lists a selection of concrete and slurry mixes to use for that project. If one of the choices is a custom recipe, you'll note that its number refers back to the mix recipes on pages 28 to 31.

You may find you already have most of what you need at hand. In which case, lay down some plastic, strap on your dust mask, put on your gloves, and get started! Remember safety first, but above all, have fun!

Simple Bowl

DESIGNER: Sherri Hunter

By using two plastic bowls for molds and a bag of premixed ingredients, it doesn't get much easier than this! Of course you can use custom recipes to cast a variety of bowls for your garden. Set your potted plants directly in them, or drill a hole in the bottom with a masonry drill bit for drainage.

Materials and Tools

Concrete Mix: Premixed Sand/Topping Mix, Mortar Mix, Commercial-Grade Mason Mix, Mix 3, Mix 4, Mix 5, or any of the Hypertufa Mixes

2 plastic bowls (1 large, 1 small)

Mold release agent

Container for mixing concrete

Small bag of sand to fill smaller bowl (rocks can be substituted)

Putty knife

Sponge

Container to hold water

Pliers

Instructions

1 Refer to instructions on pages 40 through 42.

Decorative Bowls

DESIGNER: **Kem Alexander**

Kem shares her techniques for adding interest to a simple bowl. Glistening gems or ginkgo-leaf fossils are just two of the variations. By using the basic principles, you can select your own great embellishments.

Gem Bowl

Materials and Tools

- Concrete Mix: Mix 3 or Mix 5
- Plastic bowl
- Duct tape
- Petroleum jelly
- Chicken wire
- Container for mixing concrete
- Assorted glass gems
- Turntable or lazy Susan (optional)
- Sponge
- Container for water
- Muriatic acid
- Baking soda
- Oil- or silicone-based sealer

Instructions

1 Split your plastic bowl and put it back together using duct tape on the outside of the form.

2 Using petroleum jelly as your release agent, coat the inside of the bowl form.

3 Use chicken wire to make a bowl shape that is slightly smaller than the inside of the bowl form and about 1¼ inches (3.2 cm) from the form's top. Set aside until needed.

4 Mix your concrete to a thick mud- or clay-like consistency.

5 Starting at the bottom of your bowl, place the gems with the flat side flush against the surface of the bowl. Pack the thick concrete mixture in between the glass. Then add enough concrete to just cover the gems. Start building up the sides of the bowl in the same manner by laying in about 1 inch (2.5 cm) or so of gems and concrete. Then, wait for the concrete to firm about 20 minutes for each vertical inch before laying in another inch. You want to proceed this way because if you try to build up the sides too fast, the concrete will sag with the weight of the glass. Working on a turntable or lazy Susan will make it easier to rotate your piece as you add your concrete. Continue to apply the glass and concrete to cover the inside of the bowl.

6 After the concrete has set up a bit, insert the preformed chicken-wire armature you made in step 3 so it sits approximately ⅛ inch (.3 cm) away from the glass pieces.

7 Add enough concrete to cover the armature and make a smooth surface for the inside of the bowl.

8 Start to lay in the gems and concrete for the inside of the bowl. You want to place the flat side of the glass gems so they face out. Fill the space between the gems with concrete. When the entire piece is completely built and firm, wipe off the excess concrete from the glass.

9 Let the piece sit over night before removing the duct tape to release the form. Clean the gem surfaces and smooth the edges with a damp sponge.

10 Clean the piece by dipping it in a solution of 2 parts water and 1 part muriatic acid followed by a bath of water and baking soda. The water and baking soda will neutralize the activity of the acid. Wrap the piece in wet towels and plastic for 30 days.

11 After the piece is thoroughly dried, burn off any fibers. Dip the bowl in a bath of silicone- or oil-based sealer.

Ginkgo Leaves

Materials and Tools

- Concrete Mix: Mix 5
- Plastic bowl
- Duct tape
- Ginkgo leaves
- Spray adhesive
- Spray mold release agent
- Container for mixing concrete
- Sponge
- Container for water
- Muriatic acid
- Baking soda
- Oil- or silicone-based sealer

Instructions

1 Same as step 1 for the Gem Bowl.

2 Using the spray adhesive, coat the ginkgo leaves and adhere them to the bowl with the veins towards the inside. Then, spray the bowl with a mold release agent.

4 Pack the bowl with concrete and allow it to set for about 3 hours. Carve out the interior of the bowl (see Wet Carving on page 77) and let the bowl continue to set until firm, about 3 more hours.

5 Remove the duct tape to release the form. Carefully remove the ginkgo leaves from the bowl.

6 Follow steps 10 and 11 for the Gem Bowl.

Snail Hanging

DESIGNER: Sherri Hunter

Generally, snails aren't welcome garden guests, but most people would make an exception for one like this—you can enjoy him without any threat to your plants. Of course with this technique, your wall hanging can take on any design, and each piece will be one of a kind.

Materials and Tools

- Concrete Mix: Premixed Concrete, Mortar Mix, Commercial-Grade Mason Mix, Mix 1, Mix 3, or Mix 5
- Wood for frame
- Saw
- Wood fasteners
- Reinforcing material
- 18-gauge (or heavier) galvanized or stainless steel wire for hanging loops
- Needle-nose pliers
- Insulation foam
- Cutting tool
- Paper for pattern
- Permanent marker
- Sandpaper (optional)
- Soldering iron
- Foam adhesive
- Foam tape (optional)
- Mold releases (dish soap works well on the foam)
- Container for mixing the concrete
- Colored pigment (optional)
- Trowel
- Hammer

Instructions

1 Cut the wood for the form (A and B).

2 Butt the panel ends over the panel sides and fasten them together to create a 15 x 30-inch (38.1 x 71 cm) casting frame. Set aside until needed.

3 Cut your reinforcing material to 14 x 29 inches (35.6 x 76 cm) and set aside until needed.

4 To make hangers, cut two pieces of wire, each 12 inches (30.5 cm) long. Bend the piece of wire in half using the needle-nose pliers. Bend this loop 1 inch (2.5 cm) from the top. Take the remaining end of wire and bend each side until they loop around like butterfly wings. The loops will be submerged in the concrete and the central bend in the wire will be used for hanging the panel. If you just stick two straight ends of wire into the concrete, you run the risk of having them pull out. Set them aside until needed.

5 Cut a piece of insulation foam to fit the form.

6 Follow the instructions listed on pages 45 through 47.

Cut List for the Casting Frame

Code	Description	Qty.	Materials	Dimensions
A	Panel Sides	2	1 x 4 stock	3½ x 30½ inches (8.9 x 77.5 cm)
B	Panel ends	2	1 x 4 stock	3½ x 16½ inches (8.9 x 41.9 cm)

Hypertufa Trough

DESIGNER: **Elder G. Jones**

Antique stone troughs are sought-after garden treasures that are hard to come by. Not to worry. Hypertufa vessels are both inexpensive and easy to make. These planters will bring you three times the enjoyment—when you make them, plant in them, and admire their beauty in any setting.

Instructions

1 Following the cut list, cut the plywood pieces (A and B).

2 Butt the end pieces (B) over the edges of the sides (A) to form a rectangle, and screw them together.

3 Apply a heavy mold release to the inside of the plywood form and place it on the plastic work board.

NOTE Hypertufa is a soggy mix. You may want to wrap the boards in plastic before assembly rather than using the mold release.

4 Mix your concrete and fill the bottom part of your form with at least 2½ inches (6.4 cm) of the mix. Use a block of wood to tamp or pack the mixture to eliminate air bubbles.

5 Insert the inner form, leaving about a 3- to 3½-inch (7.6 to 8.9 cm) space all the way around to ensure that you'll have enough material for shaping and carving. Continue to fill the sides, tamping as you go.

6 Trowel off the top to level, and let the casting set for 6 to 8 hours.

7 Remove the inner form first and complete any carving before removing the outer form. A putty knife will help with shaping the inside space. Take care not to carve any area too thin. Carve drainage holes in the bottom of the trough now or wait until the piece has cured and drill them with a masonry bit.

8 Unscrew and remove the form.

9 Use the carving tools to shape the outside of your trough. You may just want to round the corners. A sharp knife may be more effective to cut edges and surfaces, however a tool with a saw-toothed blade can add a good overall finished texture to the piece. Leave the piece uncovered and undisturbed overnight.

NOTE Hypertufa is very fragile at this stage.

10 The next day, spray the trough with water and cover it with plastic. Leave it undisturbed for 3 to 4 days before moving it outside. Keep it wrapped and allow it to cure for 2 weeks.

Materials and Tools

- Concrete Mix: Mix 7, Mix 8, Mix 9, or Mix 10
- Tape Measure
- Saw
- ¾-inch (1.9 cm) plywood
- Phillips head screwdriver
- Drywall screws
- Mold release agent
- Plastic work board
- Cardboard
- Pins
- Polystyrene form
- Container to mix concrete
- Block of wood
- Trowel
- Putty knife
- Assorted carving tools (see page 77)
- Lightweight plastic
- Drill with a masonry bit (optional)

NOTE: The finished trough pictured measures 10½ x 12½ x 33½ inches (26.7 x 31.8 x 85 cm).

The Inner Form

An inner form placed in the center of the casting prevents extra carving. Elder uses a collapsible box made out of wood. It has screw-on supports and crosspieces to keep the box from bowing. You could also use a block of polystyrene foam. Cut it at an angle from side to side and pin it back together. Wrap the block in plastic before putting it in place. To remove the form, you unpin the foam, slide out the sections, and remove the plastic. Another option is to tape heavy pieces of cardboard into a box that you cover with plastic. Then, put a plastic bag inside the box and fill it with sand as you fill the sides of the mold with the hypertufa mix.

Cut List for the Mold

Code	Description	Qty.	Material	Dimensions
A	Trough sides	2	¾-inch (1.9 cm) plywood	12 x 36 inches (30.5 x 91.5 cm)
B	Trough ends	2	¾-inch (1.9 cm) plywood	12 x 36 inches (30.5 x 40.6 cm)

Vegetable Steppingstones

DESIGNER: Sherri Hunter

Step into your vegetable garden in style with these reverse-cast steppingstones. Picking a theme like flowers, bugs, or butterflies is one way to create an artistic, attractive pathway. Or, you can place some of your special steppingstones among purchased ones to define a focal point.

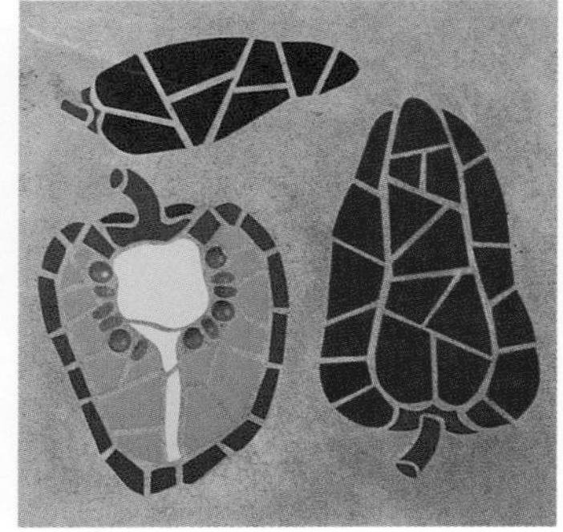

Materials and Tools

- Concrete Mix: Premixed Sand/Topping mix, Mortar Mix, Commercial-Grade Mason Mix, or Mix 3
- 5½ linear feet (1.7 m) of 1 x 3 stock*
- 1 linear foot (30.5 cm) of 2 x 2 stock*
- Drill
- Sandpaper
- Phillips-head screwdriver
- 16 drywall screws, 1¼ inches (3.2 cm) long*
- Paper for template
- Pencil
- Permanent marker
- Assorted flat, frost-resistant materials**
- Tile nippers
- Tile cutters
- Hammer
- Newspaper
- Diamond sanding block (optional)
- Clear adhesive shelf paper
- Scissors
- 2 work boards, at least 2 inches (5 cm) larger than your mold
- Mold release agent
- Hardware cloth
- Container to mix concrete
- Screed
- 4 ml plastic sheeting
- Sponge
- Container for water
- Nylon pot scrubber

*Quantities listed are for one mold. Multiply the quantity for the number of molds you're making.

**The flat side of glass gems can be used.

Instructions

1 Cut the sides (A), the ends (B), and the supports (C). Lightly sand all edges.

2 Predrill two holes at the end of each of the side pieces (A). The holes should be slightly smaller than the drywall screws.

3 Attach supports (C) to each end of the side pieces (A). See figure 1. Insert the screw into the supports from the inside.

4 Place the two sides (A) on their edges with the supports facing out. Predrill two holes at each end of the end boards (B). Line up the ends with the supports and attach them with two drywall screws, screwing into the supports to form a square. These are the screws you'll remove to take the mold apart.

5 Mark the top edges of your mold with a permanent marker so you'll know how the mold goes back together. If you've made more than one mold, put a number on each side piece so you can keep tract of your mold parts. Example: Put a number 1 on all sides of your first mold, a number 2 on all sides of your second mold and so on.

6 Follow the instructions for Reverse-Cast Mosaic on pages 102 through 105, Chapter 3.

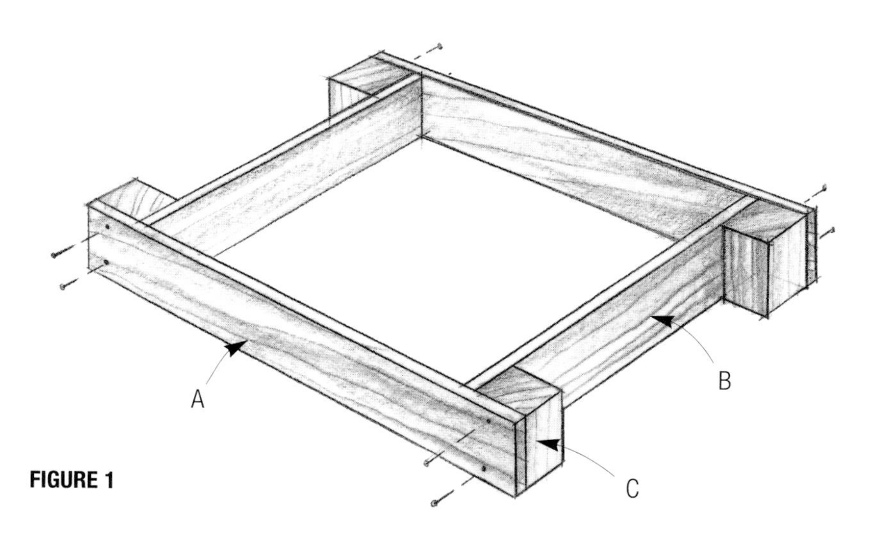

FIGURE 1

Cut List for the Mold

Code	Description	Qty.	Materials	Dimensions
A	Sides	2	1 x 3 stock	2 x 14 inches (5.1 x 35.6 cm)
B	Ends	2	1 x 3 stock	2 x 18¼ inches (5.1 x 46.3 cm)
C	Supports	4	2 x 2 stock	2 inches (5cm) long

The finished mold will cast a 2-inch (5 cm) thick, 14-inch-square (35.6 cm) square steppingstone.

Garden Mushrooms

DESIGNER: **Tom Rice**

Even if you can't grow mushrooms, you can produce a crop of these using a simple sand-casting technique. Once you start making these little mushrooms, it'll be hard to stop. That's what happened to Tom. Hundreds of these mushrooms (with no two alike) have sprouted up around his studio, adding interest to his boards and beds. Each mushroom is individually cast upside down in sand. While you're making one, you might as well make a dozen. Add even more personality to the caps by pressing a few pebbles or shells into the sand so they're picked up in the casting.

Materials and Tools

Cement Mix: Premixed Mortar Mix, Commercial-Grade Mason Mix, or Mix 3

Large plastic tub (for sand casting)

Sand

Spoon

Landscape nail*

Tarpaper

Duct tape

Container to mix concrete

Fork, kitchen knife, or other carving tools

*You can use 6- to 12-inch (15.2 to 30.5 cm) nails, ¼-inch (6 mm) metal rods or #3 rebar. These will provide the armature support and form a spike 2 to 6 inches (5 to 15.2 cm) long (depending on the size of your mushroom) which will be pushed into the ground to make your mushroom stand. To "plant" your mushrooms, make a starter or pilot hole in the soil by using a screwdriver or similar object. Avoid pushing on the mushrooms caps.

Instructions

1 Put at least 3 inches (7.6 cm) of sand in a large plastic container. Make sure the sand is damp enough to hold its shape when squeezed in your hand.

2 Make impressions into the damp sand to form the mushroom caps. You can either find a shape to push into the sand, or use a spoon to dig out the shapes you like.

3 Create the stem mold using rolled up tarpaper* and duct tape, and set aside.

4 Mix the concrete to a pourable consistency and fill the cap impressions.

5 As the concrete sets (about 30 minutes) place a nail/rod into the center of the cap and push in 1 inch (2.5 cm). Hold or brace the nail/rod in place until it stands on its own.

6 Place the stem mold so it rests on the mushroom cap with the nail centered.

7 Mix a second batch of concrete to a pourable consistency. Hold the stem mold in place and slowly fill, leaving enough of the nail/rod exposed to become the spike that plants the completed mushroom in the ground.

8 Allow the concrete to set about 12 hours.

9 Carefully remove the mushroom from the sand mold. The concrete is set but still soft enough to carve. Shape the cap using a fork or selected tool to carve the surface.

10 Remove the tarpaper stem mold and detail the stem as desired.

11 To cure, soak the mushrooms in water for seven days.

NOTE If you don't want to buy a roll of tarpaper, try visiting a housing construction site. Scraps of this roofing material can often be found in the waste pile.

Faux Boulder, Let's Rock!

DESIGNERS: **Sherri Hunter and Harry C. Kellogg**

Have you been looking for the perfect boulder for your garden? Well, here it is, and you don't need to worry about hiring a truck and a front loader to get it to its location. You can virtually build this on site. Depending on its size, you may still need some help moving it, but pound for pound, you can get a lot more boulder for your money by building it yourself.

Instructions

1 Draw a design for a basic form or just start working directly with your armature material. Estimate the base circumference of your boulder, and cut a piece of rebar about 12 inches (30.5 cm) longer. Using a bender, bend your rebar (see page 68). You can also bend the rebar in an irregular shape with the ends overlapping by using a sturdy object that will provide the necessary resistance. Wire the overlapping ends together.

2 Estimate the profile and height of your boulder. Cut a piece of rebar to fit that estimation adding a little extra length to play with. Bend an interesting curvy line so the ends will meet two opposing points on the base shape. Cut the leg of the shape if necessary.

3 Estimate the profile and height of the boulder in the direction perpendicular to the first view. Bend the rebar so that it touches the base and the first profile where they cross. Wire the two pieces together and then wire the ends to the base. Add additional pieces of rebar if you think it will help define your form.

4 Wearing your leather work gloves, use the aviator shears to cut a section of the expanded metal to cover the rebar armature. (Be careful, this stuff is sharp!) Start by folding the mesh around the base, and then using tie wires or hog rings, secure the mesh to the formed rebar. You may want to hit the mesh with a hammer to add interesting contours. (See Notes on Fastening on page 70.)

Materials and Tools

Concrete Mix: Premixed Sand/Topping Mix, Mortar Mix, Commercial-Grade Mason Mix, or Mix 3

#3 rebar

Rebar cutter

Rebar bender

Expanded metal mesh

Aviator shears

Tie wires

Tie wire winder/twister

Hog rings (optional)

Hog ring pliers (optional)

Leather work gloves

Hammer

Container to mix concrete

Flat trowel

Margin or pointing trowel

Chisel

Rock-textured stamping molds (optional)

Aluminum foil

Rub block

Sharp kitchen knife

4 ml plastic sheeting

Pressure washer

4 shades of paint for rock

Paint mixing containers

Paint mixing sticks

Disposable paint filter

Garden sprayer with metal tip

Hand-held water sprayer

Pressurized air source

Large plastic syringe

5 Once you're happy with your mesh-covered armature, mix your cement to a brownie-batter consistency. Using your flat trowel as a hawk and your margin or pointing trowel for concrete application, cover the mesh with your first coat of concrete. (Sometimes you just have to use your hands. You decide.) You will probably still see some of your mesh on this coat, and that's okay. Cover the form with plastic and let it sit undisturbed overnight.

6 The next day, uncover the form and dampen it with water. Turn the boulder over. Use a chisel to remove any large deposits of concrete that may have formed. Mix up a smaller batch of concrete. Use this batch to generously cover your rebar armature. (Rust occurs when metal is exposed to water and air. Covering the rebar with cement will help deter this.) Cover the form with plastic and let the concrete set overnight.

7 Turn the boulder right side up. Using a rub block, remove any extreme rough spots. Mix your concrete and apply a third coat. This is your final coat and will be the surface that you'll texture. While you can apply concrete to small sections at a time, make sure that ultimately the whole surface has a good coat. As you work, apply more concrete to some areas for added character. Let the concrete firm enough so it isn't sliding as you work. Experiment with rock-textured stamping molds if you're using them, or use a crumpled-up piece of aluminum foil to stamp with.

NOTE Sometimes having photographs or a small rock as a model helps you to understand the variations of texture that can be found on one rock.

8 As the concrete gets firmer, use the point of a knife to create ridges, layers of striation, and crevices. You may even want to create a fossil or two by stamping the surface with a shell or by pressing in a leaf. At this point, you are the sculptor. Remember to work over the whole boulder; don't get too detailed in one small area. Work the surface until you are satisfied, or until the concrete is too hard. Cover with plastic and leave undisturbed overnight.

9 The next day use the pressure washer to thoroughly clean your boulder of all debris. Let the piece dry completely before painting it.

10 Follow the instructions in Paint as Stain on page 109 to finish your boulder.

Tribal Mask

DESIGNER: **Sherri Hunter**

Were you going for the tropical-jungle feeling when you planted your lush garden? This over-sized, stylized mask will provide the perfect focal point. Not only will it endure the elements, it'll keep a watchful eye on everything in your absence. What's even better, the sand-casting technique used to create this mask can be adapted for other garden projects, including bowls, birdbaths, and fountains.

Instructions

1 Follow the instructions for Sand Casting, pages 48 to 49, steps 1 through 6.

2 This mask is designed to hang on the wall. The easiest way to do this is to insert hanging loops into the wet concrete as you're making the mask. Cut two pieces of wire 10 inches (25.4 cm) long. Bend them in half using your needle-nose pliers, forming a narrow loop. Bend them again 1 inch (2.5 cm) from the top of the loop. Twist the ends of the wire into overlapping circles so the 1-inch loop stands up perpendicular to the overlapping circles (refer to photo 43 on page 49).

3 Position the loops through the plastic, about 2 inches (5 cm) in from the edge of your sand mound, and at a distance about one-quarter the length of your mask from the top. Remember, the sand represents the open space behind the mask. You want the loops to be sticking out of the concrete surface.

4 Continue to follow the Sand Casting instructions, steps 9 through 12.

5 Pull the edges of the thin plastic toward the center of the casting. This will help to coax the concrete back—these things have a way of growing as you work—and compress the material at the edge, making it stronger. Leave the plastic in the drawn-up position.

6 Continue to follow the Sand Casting instructions, steps 14 through 17.

7 You now have the base for your mask. Make a mark where you want to place the mouth and eyes. Follow the instructions for Texturing a Mask, page 90, steps 2 through 6.

8 If you're going to paint the mask, continue to cure it for about three more weeks. Refer to Paint as Stain, pages 109 through 110, and read any surface preparation information on the label of the paint you've selected. In this project, the entire piece was painted front and back with two coats of an exterior blue paint and allowed to dry. Next, a gloss-black paint was brushed over the surface and then rubbed off with a rag, leaving more of the black paint in some of the areas to better define the features.

Materials and Tools

- Concrete Mix: Premixed Sand/Topping Mix, Mortar Mix, Commercial-Grade Mason Mix, Mix 3, Mix 5, or Mix 12
- Slurry Mix: Mix 20, Mix 21
- 4 ml plastic sheeting
- Approximately 2 gallons (7.6 L) of sand
- Plastic tub
- Water
- Thin plastic bags or plastic drop cloth
- Permanent marker (optional)
- Aviator shears
- Reinforcing material
- 18- to14-gauge galvanized or stainless steel wire
- Angle wire cutters
- Needle-nose pliers
- Container to mix concrete
- Putty knife
- Brush
- Large plastic trash bag
- Course rasp or file
- Masonry drill Bits
- Drill
- 4½ inch (11.4 cm) grinder with masonry wheel
- Container to mix slurry
- Paintbrush
- Incising tools
- File or wire brush
- Blue and black exterior gloss paint (optional)
- Heavy-duty picture hanging wire

Garden Bench

DESIGNER: Sherri Hunter

This is a perfect place to perch and enjoy a sunny afternoon—and it's easy enough to make in a day. I used two plastic tote bags as molds for the legs (look for the yellow tote in the group photo of molds on page 34), while the shape for the seat was cut out of polystyrene insulation. I used white Portland cement and decorative rocks that I found at my local home improvement center as the aggregate. The beauty of this project is that you can use any combination of material to complement your setting.

Materials and Tools

Concrete Mix: Premixed Concrete, Mix 1, or Mix 2

2-inch-thick Polystyrene (bead board)

Paper for seat template

Scissors

Tool to cut polystyrene

Long nails (optional)

Duct tape (optional)

8 long screws

8 large washers*

4 ml plastic

Two molds for legs

Mold release agent

#4 Rebar

Tool to cut rebar

Container for mixing concrete

Block of wood

Trowel

Stiff nylon or wire brush

Spray water bottle

Grinder with masonry wheel (optional)

Silicone adhesive (optional)

* These are used to keep the screw heads from sinking into the polystyrene. Because they're only used temporarily in the mold-making process, I improvise and use the plastic tops from milk containers.

Instructions

1 Draw the shape of your seat on the paper. Cut out the shape and use it as a pattern on your polystyrene foam. Your foam should be at least 2 inches (5 cm) thick. If you have thinner sheets of foam, pin them together using long nails. Position your pattern on the foam board so that it is at least 1½ inches (3.8 cm) from any one side. Trace your pattern and cut the shape out of the foam. Because the negative space will be your mold, don't worry if you need to remove the center part of the foam in pieces. You may even want to cut your foam into two pieces to make cutting out the shape easier.

2 If you cut your foam into two pieces, you may want to wrap duct tape around the edges to hold the sections together. Secure you polystyrene mold to a level, plastic-covered work surface with the long screws and washers, as shown in figure 1.

3 To reinforce your seat, cut three lengths of rebar 3 inches (7.6 cm) shorter than the length of your seat. For the bench in the photograph, each piece of rebar was cut at a different length because of the curved shape. Set the cut rebar aside until needed.

4 Select your molds for the legs and generously apply mold release.

5 Mix your concrete to a firm, clay-like consistency. It should still hold together after you squeeze it in your hand. It will probably take more than one batch of concrete to fill all three of the molds.

6 Begin to fill the seat mold. Use the block of wood to tamp the concrete, paying particular attention to any corners and edges. Fill the mold halfway, and then lay in the pieces of cut rebar so they're equally spaced and do not touch the walls of the mold. Continue to fill the mold to the top, tamping as you go. Level with a trowel, but do not overwork the surface or the decorative aggregate will sink lower into the concrete.

7 Fill your leg molds, and tamp them with the block of wood, paying the same attention to the edges as you did with the seat. You can fill these forms to be solid, or you can insert a filler piece of polystyrene into the center after you have cast at least 3 inches (7.6 cm) into the mold. The filler must allow for a space of at least 3 inches from the sides of the filler to the mold. Continue to cast the mold until it's full and then level it.

8 Let the pieces set until firm, about 6 to 8 hours. You should be able to rub your finger across the surface without any water appearing, but should still be able to scratch the concrete with your fingernail.

9 Remove the screws, duct tape, and the polystyrene from your bench seat, breaking the polystyrene mold as needed. Demold the legs by carefully turning them upside down. Clean the pieces while they are still in this position.

10 Using a hose or a handheld sprayer, mist the surface and scrub it with a stiff brush to remove the concrete coating, which will expose the aggregate. Spray with water frequently during this process to wash away the concrete. If the concrete has set too long, you may want to use a grinder with a masonry wheel to reveal more of the aggregate. A muriatic acid and water wash will help remove the additional concrete haze.

11 Cover the pieces in plastic and allow them to cure for 5 days or longer before assembling your bench. Gravity will keep the heavy pieces in place. If you're concerned that they might move, use a silicone adhesive to hold them in place.

NOTE The dimensions for the bench shown in the photograph are:

LEGS/BASES

8 x 9½ x 16 inches
(20.3 x 24.1 x 40.6 cm)

TOP

3 x 44 x 16 to 19 inches
(7.6 x 111.8 x 40.6 to 48.3 cm)

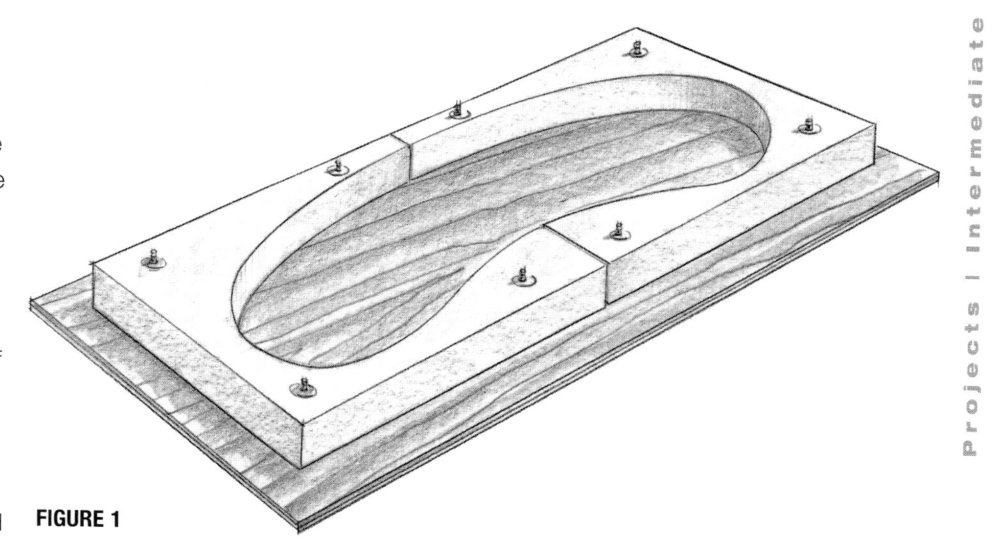

FIGURE 1

Leaf Bowl

DESIGNERS: Marsha Hoge and Dixie Stephens

Want to make a *big* statement in your garden? Use an elephant ear leaf to create this elegant bowl. With the support of a mound of sand, you can cast any variety of leaf to capture its natural beauty. Then you can accentuate it by using various coloring techniques. Use these bowl as birdbaths, garden accents, or even rainspouts.

Instructions

1 Refer to Sand Casting, page 48, steps 1 through 3, and shape an elongated oval with your sand.

2 Cover the sand with a layer of thin plastic.

3 Cut off the leaf stem and place the elephant ear leaf face down on the mound.

4 Mix your concrete to a muffin-batter (thick) consistency.

5 Take a handful of concrete and form it into a ¾-inch-thick (1.9 cm) "patty." Lay it over the stem-nub portion of your leaf. Continue to make patties and apply them, first along the central vein of the leaf and then down the sides. Overlap the patties as you work, patting them slightly so the seam disappears.

6 Build up the edge of your bowl just shy of the edge of the leaf, adding extra concrete to form a nice smooth surface. Avoid tapering the concrete; that would result in fragile edges.

7 Smooth the edges and back of your bowl with your hands or with a moist paintbrush. Cover with plastic and let it set overnight.

8 The next day, tap the concrete to be sure it has hardened. Pick up the bowl, along with the plastic that was covering the sand. Use the sand on your work surface to cushion the bowl as you turn it over and set it down.

9 Remove the plastic and the elephant ear leaf. Don't worry if not all of the leaf comes off at this time. Gently remove any rough edges with a rasp.

10 Wrap the bowl in plastic and allow it to cure for 5 days. After that, if some of the leaf remains in the cast, you can use a nylon brush to scrub the surface to remove the plant residue.

11 Allow the piece to dry thoroughly before treating it with concrete sealer.

Materials and Tools

Concrete Mix: Premixed Mortar Mix, or Mix 3*

4 ml plastic sheeting

Sand

Water

Thin plastic drop cloths

Elephant ear leaves

Utility knife

Container to mix concrete

Paintbrush fro smoothing edges (optional)

Pigments for coloring the concrete (optional)

Rasp

Stiff nylon brush

Concrete sealer (optional)

*Use a 50/50 polymer/water combination instead of plain water for any mix you use.

Flower Power

DESIGNER: Ricky Boscarino

Ricky started small when he began working on garden sculptures, but things just kept growing. Today there's Luna Parc, the art environment he's created in New Jersey.*

Instructions

1 Design your flower and leaf shape and make paper templates from the design. Using the templates, trace the design on the plywood. If you prefer, you can draw your designs directly on the wood. Make sure that no sections of your designs are too narrow or they will become weak points in the casting. Cut out the plywood shape.

2 Cut the roof flashing into long 3¾-inch-wide (9.5 cm) strips. Using the drywall screws, attach the roof flashing along the edge of the wood. Space the screws every 2 to 3 inches (5 to 7.6 cm) so there are no gaps. You want the metal to remain perpendicular to the wood and flush to the backside. Overlap the metal at least 4 inches (10 cm) as you completely enclose the shape. Tape the outside overlap with duct tape. Do this to both the flower and the leaf shape.

3 Measure the spaces in your mold and cut strips of reinforcing mesh that will fill in the shapes. Set aside until needed.

4 To locate the stem of the flower, mark the center of the metal from the inside of the mold and drill a hole using the spade drill bit. Drill two holes, opposite of each other, where the stem will be located on the leaf shape.

5 Lay the flower and leaf form on the ground in the position you want them. Thread a piece of rebar through the leaf and into the flower 2 inches (5 cm) from the roof flashing. It's better if the rebar extends into the petal to give you added reinforcement. Determine how long of a stem you want and add 10 inches (25.4 cm) more. This extra length will be the section that anchors the flower in the ground. Mark and cut the rebar to this length.

6 Apply mold release to your molds and position them on a work surface for casting. Insert the rebar. Mix your concrete to a batter-like consistency.

7 Fill the molds with 1 inch (2.5 cm) of concrete. Insert several strips of your reinforcing mesh so they go under the rebar. Continue to fill the mold another inch and lay in the remaining strips of reinforcing mesh. Continue to fill the mold. Tap the sides of the mold lightly to remove air bubbles, and level the casting with a trowel.

8 After the concrete has begun to firm, embed your selected materials by pushing them into the wet concrete (see Embedding on a Base Form, page 106). Cover with plastic and allow the form to cure one week before removing from the mold.

Materials and Tools

- Concrete Mix: Premixed Mortar Mix, Commercial-Grade Mason Mix, or Mix 3
- Paper for making templates
- Permanent marker
- ¾-inch (1.9 cm) plywood
- Saw
- Aluminum roof flashing
- Scissors
- 1-inch (2.5 cm) drywall screws
- Drill with Phillips-head attachment
- Duct tape
- ⅝-inch (1.6 cm) spade drill bit
- #4 rebar
- Rebar cutter
- Tape measure
- Reinforcing mesh
- Mold release
- Container to mix concrete
- Assorted frost-resistant materials
- Found objects

9 Carefully remove the roof flashing. Clean the embedded pieces of any stray concrete. Pick up the cast piece by its edge to remove the plywood. You may need help to lift the form. File the edges as needed.

10 To plant your flower, push the stem into the ground.

*You can visit it by going to www.lunaparc.com.

Birdbath

DESIGNER: **Sherri Hunter**

While a birdbath helps you care for the birds in your garden, you should consider your needs as well. You'll want the birdbath to be in a location where you can enjoy observing the birds as well as admire your beautiful one-of-a-kind creation. Before you start this project, plan where you'd like to place the birdbath. You want to take the location into account as you determine the height of your base and design the bowl.

Instructions

1 Design your bowl and construct it following the instructions for Sand Casting, pages 48 to 50, steps 1 through 6, and 9 through 12. Be careful not to make your bowl too deep—you want to make it attractive for songbirds, not ducks.

2 Follow the instructions for the Bowl Connector, page 52, steps 2 through 5, to finish your initial form.

3 To make your pedestal base, follow the instructions for Simple Column Armature, pages 63 to 65, steps 1 through 14, taking into consideration the Column Connection instructions, page 66, steps 1 through 4. It helps to balance a design if the foot of your pedestal relates to the shape of your bowl. If you have a square bowl and would like a square foot on your pedestal, follow the instructions for Making a Square Foot for a Column, pages 65 and 66, steps 1 through 5.

4 The final layer of concrete for this project was finished using the embedding process. Follow the instructions for embedding, pages 107 to 108. Your actual design can be anything from flowers to abstracts.

5 Install your birdbath on a level spot. I like to place the base on a pre-cast steppingstone that has a diameter slightly larger then the foot. That way, the decoration doesn't get lost in the grass or hit by the lawnmower. Insert the stem of the bowl into the space at the top of the base. Fill with water and wait for the birds. Don't be surprised if butterflies stop by too. Change the water regularly and scrub with a nylon brush to keep clean.

Materials and Tools

Concrete Mix: Premixed Mortar Mix, Commercial-Grade Masons Mix, or Mix 3

4 ml plastic sheeting

Approximately 2 gallons (7.6 L) sand

Water

Dry-cleaning bags or thin plastic drop cloths

Permanent marker

Aviator shears

Reinforcing material

Container to mix concrete

Two 16-ounce (.47 L) plastic cups

Large plastic trash bags

Coarse file or rasp

4-inch (10 cm) diameter cardboard or plastic tube

Tape measure

Metal mesh, ½-inch (1.3 cm) hardware cloth

Angle cutters

6-inch-long (15.2 cm) pieces of 22-gauge or plastic-coated wire

Leather work gloves

Chicken wire, 1-inch (2.5 cm)

Hammer

Putty knife with tooth-shaped blade

Container to mix slurry

Paintbrush

Assorted frost resistant materials and/or found objects

Construction Timeline

Depending on the time you have available, making a birdbath can be a long-term project. In the workshops I teach, though it's rather intense, we construct the elements in one weekend in the following sequence.

Friday Evening

Cast the bowl.

Saturday Morning

Demold the bowl and file the edges.

Make the base armature and coat with first layer of concrete.

Saturday Afternoon

Add a layer of concrete and embed the inside of the bowl.

Sunday Morning

Clean the embedded bowl.

File the edges of the bases.

Start adding concrete and embedding the base.

Sunday Afternoon

Finish the bases.

Finish the backs of the bowls.

Monday Morning (with the help of my studio assistants)

File any edges and clean embedded areas.

When the Workshop Participants Can

Clean birdbaths with muriatic acid and take them home to enjoy.

This is a very rigorous production schedule, but by doing the pieces in this order, the concrete has ample time to set firmly before moving on to the next step. You can easily approach this project in shorter work sessions to complete the birdbath over a couple of weeks.

Who's Chicken?

DESIGNER: **Virginia Bullman**

Not this plucky bird! This is definitely a chicken with attitude. Because one of Virginia's daughters raises chickens, she's had the chance to closely observe these animals. When you make art from sources you know well, as Virginia has done here, you can capture the essence of your subject and have fun, too.

Instructions

1 Referring to the chart on page 148, cut the rebar for the legs and slightly bend the leg sections (A) 4 inches (10.2 cm) from one end,. Cut and bend the rebar for the body (B) and bend it into a slight elipse that measures 6 inches (15.2 cm) across by 7 inches (17.8 cm) long. Weld the rebar sections together to make the armature, referring to photo 93 on page 67.

2 Select or construct a mold for casting the base of your armature. Mix your concrete for the base and fill the mold. Allow the concrete to set up until stiff enough, and then insert the legs into the center of the form. Brace the armature if needed, but if it's slightly angled it will have more "chicken attitude" (refer to photo 95 on page 67). Allow the form to cure overnight.

3 Take the chicken wire (C) and wrap it around the rebar armature. You will need to cut slits in the mesh where it wraps around the legs so the mesh can bend and overlap underneath the rebar. Secure the mesh to the frame with a few pieces of wire.

4 Allow more of the chicken wire for the head and neck than the tail. Overlap the head sections to form a cone, and secure with wire. Wearing gloves, squeeze and twist the mesh to form the neck and head. Bend and pose. Remember, chickens are all body and little brain; their tails are small too.

5 Before closing the tail, wad up sheets of newspaper and stuff them into the body cavity. Squeeze the top and bottom of the tail together until it's almost flat at the end, and fold in the corners to form a triangle (refer to photo 94 on page 67).

6 Shape the small pieces of chicken wire (D) into cones, and place them around the upper part of the legs. Wire the cones on the outer sides of the the body.

7 Mix the concrete to a clay-like consistency. Make sure you thoroughly mix the dry ingredients together before adding water.

Materials and Tools

- Concrete Mix: (Base) Premixed Mortar Mix, Commercial-Grade Masons Mix, or Mix 3; (Chicken) Mix 8
- Rebar
- Rebar cutter and bender
- Welder
- Mold for casting base
- Concrete
- Container for mixing concrete
- Chicken wire, 1 inch (2.5 cm)
- Leather work gloves
- 22-gauge wire
- Newspapers
- Assorted mosaic materials (see sidebar)
- Beads for eyes (see sidebar)
- Grout
- Sealer
- Paintbrush
- Yellow paint

8 Start applying concrete to the top of the body, the sides, the neck, and head (refer to photo 95 on page 67). When the concrete starts to fall off, stop, cover the form with plastic, and give the concrete time to set up.

9 The next day, mix a fresh batch of concrete and apply it to the bottom and upper legs. Cover and let the concrete set up.

10 For the final session, add feet and a finishing coat to smooth the body and define the details of the chicken. Cover the form with plastic and allow it to cure for two weeks.

11 Allow the chicken to dry completely before proceeding with the mosaic. Refer to the information on Adhering Mosaic, pages 99 to 100, or follow Virginia's methods found in the box on the right.

12 Paint the remaining exposed rebar leg with yellow paint. The pattern of the rebar actually looks like the texture of a chicken leg. After the piece is cured and dried, brush on a coat of sealer. Additional coats of sealer will add more shine.

Cut List for the Armature

Code	Description	Quantity	Materials	Lengths
A	Legs	2	Rebar	10 inches (25.4 cm)
B	Body base	1	Rebar	21 inches (53.3 cm)
C	Body	1	Chicken wire	18 x 24 inches (45.7 x 61 cm)
D	Drumsticks	2	Chicken wire	4 x 4 inches (10.2 x 10.2 cm)

NOTE: Refer to the information on Simple Animal Armature on page 000.

How Virginia Bullman Mosaics a Chicken

"Cut up flowerpots into triangles and diamonds (feather shapes) with nippers. Save all the little shards for the in-between places. Collect some shards of brown or gold dishes for the feathers, details, and wing outlines. For this chicken, I used unglazed porcelain tile for waddles. The eyes are hematite beads set in the bottom piece of a flowerpot. Use yellow pottery with texture for the beak and feet.

Starting with the back, bottom, and tail, set the mosaic pieces on the body in a direction like feathers grow. Save the head and feet for last.

I set the mosaic with a mix of Portland cement and acrylic bonding admix. Usually, I allow the mix to squish up around each piece and then smooth it to cover all the edges. This time I left the cracks (the spaces between the mosaic pieces) open, except on the feet and head, and later grouted it with a dark terra cotta grout. Be sure to clean the setting material off the surface of the pieces and seal them with grout and tile sealer or concrete sealer before grouting."

Zigzag Planter

DESIGNER: **Elder G. Jones**

Do you like designs that combine curves and angles? If so, this zigzag planter is the perfect project for you. The wet-carved concrete techniques used for this design can be applied to almost any bas-relief pattern you can imagine. Add color pigments to your mix to further customize your one-of-a-kind carving.

Instructions

1 Roll the galvanized sheet metal to the largest outside measurement of your planter and clamp it. Use the thin rope to secure the bottom portion of the mold.

2 Follow the instruction for Wet Carving on pages 77 through 80. On step 5 of the instructions add this consideration: Your plastic container will need to be 5 to 6 inches (12.7 to 15.2 cm) smaller in diameter than your sheet metal form. Be sure to cast your walls thick enough on your planter form to allow plenty of material for carving.

Materials and Tools

- Concrete Mix: Premixed Sand/Topping Mix*, Mortar Mix*, Commercial-Grade Mason Mix*, or Mix 4
- Sheet of flexible metal
- Small C-clamp
- Thin rope
- Sheet of plastic for work board
- Turntable or lazy Susan
- Container for mixing concrete
- Plastic container
- Sand
- Block of wood
- Pointing trowel
- Awl, large nail or other sharp pointed tools
- Sharp kitchen knives
- Assorted toothed and flat tools
- Whiskbroom

*If using a premixed concrete for wet carving, sift contents of bag before mixing with water.

Polished Side Table

Materials and Tools

Concrete Mix: Mix 2

Wood for mold*

Drill

Sandpaper

Phillips-head screwdriver

16 drywall screws, 1¼ inches (3.2 cm) long

Mold release agent

Hardware cloth

Aviator shears

Container to mix concrete

Screed

Trowel

Plastic-laminated work board

Water grinder

Rigid backer pad (backed with hook-and-loop material)

Assorted grit pads 60, 150, 300, 500, 1000 (backed with hook-and-loop material)

Skid-proof mat

Squeegee

Rubber apron (optional, but think about it)

Rubber-soled shoes

Sealer (optional)

Wax (optional)

* This table is a 2-inch-thick (5 cm), 18-inch (45.7 cm) square. The frame was constructed using the same method as the steppingstone mold illustrated on page 129. Making your mold out of a plastic-laminated wood will result in smoother surfaces even before you start your polishing process.

DESIGNER: Sherri Hunter

This smooth-surfaced accent table does equally well whether used inside or out. Stained glass scraps are the decorative aggregate in the white Portland cement. The glass pieces were tumbled in a concrete mixer with sand and water to soften their edges before adding them to the mix. The perfect base for the table is actually a stand for a hot water heater that was purchased from a home improvement center.

Instructions

1 Construct your mold. You may want to select your base in advance of casting the top to insure that they'll work together.

2 Follow the instructions for Casting a Simple Mold, pages 38 to 39, steps 2 through 7.

3 After casting, keep the form damp and cover it with plastic. Allow it to cure for about 5 days before removing the mold. The surface that was cast on the plastic-laminated work board will be the top of your table. Dampen and cover the form again to continue allowing it to cure for about 9 more days before you begin your grinding.

4 Follow the instructions for Wet Polishing with a Grinder, page 93, steps 1 through 6.

5 Take the grinding process through your 500-grit disc. Your tabletop should be feeling pretty smooth by now. If you still think it feels rougher than it should for the amount of work you've put into it, consider the information in Cement Paste Backfill, page 94.

6 Polishing actually starts where grinding ends. Let your tabletop continue to harden by curing at least a week more before proceeding with your 1000-grit disc. Additional discs at higher grit numbers can be used if desired. Allow the piece to dry thoroughly before sealing and waxing.

Butterfly Garden

DESIGNER: **Sherri Hunter**

Even if you can't attract butterflies to your garden, you can make some of your own. Carved polystyrene cores are layered with concrete and then covered with mosaic designs to provide year-round enjoyment. However, these butterflies need some help flying. They measure approximately 18 x 30 x 4½ inches (45.7 x 76.2 x 11.4 cm) when complete. By mounting them on poles, you'll help them appear to be fluttering just above your bushes.

Instructions

1 Design your butterfly. (I found that looking through nature books was a big help.) Carve your polystyrene following the instructions for Carving a Polystyrene Foam Armature, pages 57 to 58, steps 1 through 7.

Materials and Tools

- Concrete Mix: Mix 12
- Slurry Mix: Mix 21
- Polystyrene foam
- Pencil
- Permanent marker
- Cutting tools
- Assorted wire brushes
- Straightedge
- PVC pipe, 1-inch (2.5 cm) inside diameter
- Machine screw and nut
- Drill and bit
- Foam adhesive
- Large construction nails
- Lightweight AR glass scrim
- Scissors
- Container for mixing concrete
- Container for mixing slurry
- Paintbrush
- 4 ml plastic sheeting
- Galvanized roofing nails (if glass scrim is not self adhesive)
- 2 ml plastic sheeting
- Rub block or rasp
- Assorted frost-resistant mosaic materials
- Fortified thin-set adhesive
- Tile nippers
- Hammer
- Newspaper
- Grout
- Sponge
- Terry cloth rags
- PVC pipe for installation with ⅞-inch (2.2 cm) outside diameter

2 Insert a piece of PVC pipe into the form following the instructions for Adapting a Polystyrene Foam Core as a Pole Topper, pages 60 to 61, steps 1 through 6.

3 Cover your polystyrene core with a polymer fortified concrete and alkali resistant mesh system using the instructions on pages 82 to 83, steps 1 through 10.

4 With your basic form completed, you are ready to apply the mosaic design. Sketch your design onto the form using a pencil. When you're satisfied with the design, clarify the final shapes by outlining them in permanent maker. Review the information on Mosaics, pages 99 to 100.

5 Once the mosaic adhesive has dried at least 24 hours you can follow the instructions for Grouting a Mosaic Surface, pages 101 to 102, steps 1 through 7.

6 To install the butterflies, you'll need to insert the installation pole into a length of PVC pipe that is the same diameter as the piece used in the butterfly. I wanted to make bases that could be moved to different locations. However, if you wanted to install your forms in a more permanent location, you could dig a hole in the ground, center your pipe, and pour in concrete while making sure your pipe remains plumb. If you want to make a sturdy base that you can move, use the following instructions.

Movable (but heavy) Base

Instructions

1 Cut a 10-inch (25.4 cm) length of the cardboard casting tube and place it on 4 ml plastic sheeting to cast.

2 Cut your PVC pipe, and tape one end closed with the duct tape.

3 Mix your concrete.

4 Position your pipe with the closed end down in the center of the cardboard form. Start filling the form with concrete, tamping as you go and making sure that the pipe stays plumb.

5 Fill the form to the top and level with a trowel. Let the form set undisturbed overnight. The next day, you can peel the cardboard form off the concrete by pulling it along the spiraled seam. Cut the pipe flush with the top of the concrete.

Materials and Tools

- Concrete Mix: Premixed Concrete or Mix 1
- 12-inch-long (30.5 cm) cardboard casting tube
- Saw
- PVC pipe, 1-inch (2.5 cm) inside diameter and 12 inches (30.5 cm) long
- Duct tape
- Container to mix concrete
- Trowel

Bubbling Fountain

DESIGNER: **Sherri Hunter**

Every garden seems to come alive with the sound of water. This bubbling fountain, whether it's on your deck or in your favorite outside room, will add to your tranquil moments. Using simple forming techniques, in combination with an understanding of fountain basics, you can make these little water features for all of your favorite spots.

Instructions

1 Start by referring to the sidebar on Simple Fountain Construction on page 84.

2 Cut polystyrene ends (A), sides (B), and bottom (C). Use the adhesive to glue them together to make a five-sided box. Pin the ends together with the large construction nails to add strength at the corners. In the same way, use the nails to attach the bottom to the sides.

3 Carve the sphere as described on page 84, and add the copper tube into the form. Put masking tape on the ends of the copper tube before applying the concrete.

4 Following steps 1 through 6 for Polymer Fortified Concrete and Alkali Resistant Fiberglass Mesh, pages 82 to 83, apply the lightweight AR scrim to both the sphere and the water reservoir. Then continue with the same instructions, steps 7 through 10, for the reservoir only.

5 To finish the sphere add color to the concrete mix and slurry. Apply the slurry, then cover the sphere with a generous coating of concrete so that you can embed the glass gems. Make sure that your copper tube is ¼ inch (.6 cm) above the finished concrete. Here I used a 12-pointed design to place the gems, with six gems on six

Cut List

Code	Description	Quantity	Sizes
A	Ends	2	4½ x 19½ x 2 inches (11.4 x 49.5 x 5 cm)
B	Sides	2	4½ x 15½ x 2 inches (11.4 x 39.4 x 5 cm)
C	Bottom	1	19½ x 19½ x 2 inches (49.5 x 49.5 x 5 cm)
D	Sphere	1	10-inch cube (25.4 cm)

NOTE: The base of the finished fountain measures 7 x 20 x 20 inches (17.8 x 50.8 x 50.8 cm) with 2½-inch-thick (13 cm) walls. This includes the concrete and mesh application combined with the tile work. The bubbling sphere is approximately 10 inches (25.4 cm) in diameter.

Materials and Tools

- Concrete Mix: Mix 12
- Slurry Mix: Mix 21
- Polystyrene foam
- Tape measure
- Permanent marker
- Polystyrene cutting tool
- Lightweight AR glass scrim
- Heavyweight AR glass scrim
- Scissors
- Foam adhesive
- Large construction nails
- PVC pipe
- Container to mix the concrete
- Container to mix the slurry
- Paintbrush
- Assorted frost-resistant tiles and glass gems
- Waterproof adhesive
- Grout
- Concrete colorant
- Pump
- Copper tubing
- Pipe cutter
- Plastic tubing
- Round galvanized vent collar

of the points, four gems on the points between, with one gem between all points. Use your slurry brush to smooth the surface. Clean the glass gems after the concrete is firm.

6 To finish the reservoir, plan your mosaic and prepare your tile. I used six colors of tiles cut into 1⅝-inch (4 cm) squares applied randomly. Adhere your mosaic using the waterproof adhesive. Allow it to dry overnight before grouting. See instructions on grouting, pages 101 to 102 to finish your reservoir.

7 To assemble your fountain, place your cured, grouted reservoir on a level location with the pipe towards the back. Cut a short piece of plastic tube to connect the copper pipe to the pump. Place the galvanized collar with the narrow opening down in the center of the reservoir. Place the pump in the center with the cord draped over the edge of the collar. Balance the sphere in the collar opening. Thread the pump plug and cord through the pipe. Fill the reservoir within 1 inch (2.5 cm) of its top before plugging in the pump. (Running a pump dry will cause it to burn out.) Adjust the height of the water coming out of the copper tube as needed or shift the sphere so the water is coming out straight.

8 Sit back and enjoy another successful project!

Bottle Panel

DESIGNER: **Sherri Hunter**

Austrian born artist Friedensreich Hundertwasser provided the inspiration for this project. I was intrigued when I saw photographs of a public project he had designed in New Zealand and began experimenting with the possibilities of using bottles in cast concrete. Ricky Boscarino, page 46, has also found inspiration in this construction method and has added the bottle elements to some of his freestanding sculptures. You can use a panel as a stand-alone feature or incorporate it into a garden wall. Either way, you'll enjoy the sunlight as it streams through the colored glass.

Materials and Tools

- Concrete Mix: Premixed Mortar Mix, Commercial-Grade Masons Mix, or Mix 3
- Wood for the casting frame
- 10 glass bottles of different colors and shapes
- Corks
- Drill and bit
- 8 drywall screws, 2 inches (5 cm) long
- Phillips-head screwdriver
- Mold release agent
- Sand
- 4 ml plastic sheeting
- Large spoon
- Container to mix concrete
- Trowel
- Kitchen knife
- Sponge
- Nylon scrub brush

Instructions

1 Prepare your bottles by washing them, removing the labels, and allowing them to dry. After they dry, seal the bottles by inserting corks until they're firmly in place. Trim off any excess cork flush to the opening. If bottles have a screw top, tighten the tops.

2 Cut your boards for the casting frame following the cut list above. Pre-drill 2 holes on each end (B) and screw the sides (A) with drywall screws to form the frame. Measure and draw a line 1 inch (2.5 cm) from the bottom around the inside of the frame.

3 Apply your mold release agent. Place your frame on a plastic covered work surface and begin to fill it with damp sand to the 1-inch mark. Pack the sand firmly and then level it.

4 Place the bottles on the sand and press them in slightly. Lift each bottle and use the spoon to dig the indented areas a little deeper. Make sure that when in place, the neck of the bottle is at least ½ inch (1.3 cm) above the sand. Take some additional sand and pack it next to the bottle and under the neck so that it slopes slightly (see figure 1). Mist the sand with water, drying the bottles if any water gets on them. (You want the sand to be wet enough so it doesn't pull the water out of the concrete during casting, and you don't want any extra moisture on the bottles to be added to the concrete.)

5 Mix your concrete to a soft-batter consistency.

6 Start by placing and tamping the concrete with your fingers next to the necks of the bottles, making sure that the end of the neck is completely encased. Continue adding concrete around all of the bottles, paying attention to the frame's edges and corners. Fill the frame until the bottle necks are covered.

7 Smooth the concrete with a trowel; allow the bleed water to evap-

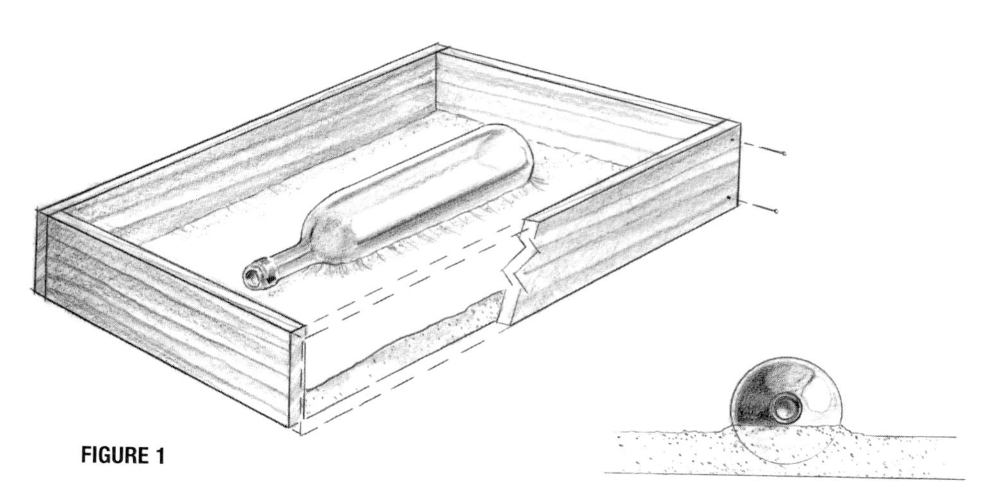

FIGURE 1

orate and the concrete to set. Don't worry if there's concrete on the bottles.

8 After about 5 hours, or when the concrete feels firm, use a kitchen knife to carve out a beveled area in the concrete to expose the body and part of the neck of the bottles. Use your sponge to clean off the bottles and smooth the concrete. Cover with plastic and let the panel remain in the frame for 2 days.

9 Unscrew and remove the wood frame. Lift up the panel and brush off as much of the sand as you can. Take the panel outside and wash off the remaining sand using the hose and nylon scrub brush. Continue to cure for 5 more days.

Cut List for the Casting Frame

Code	Description	Quantity	Stock	Lengths
A	Sides	2	1 x 6 decking stock	24 inches (61 cm)
B	Ends	2	1 x 6 decking stock	26 inches (66 cm)

Puzzling Patio

DESIGNER: **Elder G. Jones**

Arrange these whimsical steppingstones however you want. It looks like all the pieces should fit together, but in fact they don't. Once your molds are made and you've refined your carving techniques, you'll be on your way to making this project. You could even add a puzzling pathway to get to your puzzling patio.

Instructions

1 Design your puzzle shape on paper to make a template. Use the paper template as a guide to bend the metal into the desired shape. Overlap the ends to close the form and secure them with a small C-clamp.

2 Cut two or three pieces of rebar to fit into the mold. Make sure the rebar lays in all the projections of the puzzle piece (see the puzzle mold in the photo on page 34) but is at least 1 inch (2.5 cm) away from the sides of the mold.

3 Place the plastic work board on the lazy Susan and position the mold. Mix your concrete.

5 Fill the mold halfway and then insert your rebar. Make sure that when you place the rebar it remains 1 inch (2.5 cm) from the edge of the mold.

Continue to fill the mold, tamping the concrete and tapping the mold sides to remove air bubbles. Level the concrete with the trowel. Let the mold set until the concrete is firm but not hard. If the temperature is 70° to 75° F (21° to 24° C), this will probably take about 3 to 4 hours.

6 Remove the mold and shape the edges of your puzzle piece by referring to the instructions for Wet Carving, pages 77 to 80, and finishing with steps 15 through 17.

Materials and Tools

Concrete Mix: Premixed Sand/Topping Mix*, Mortar Mix*, Commercial-Grade Masons Mix*, Mix 4, or Mix 5

Paper for template

3-inch-long (7.6 cm) strips of steel**

Small C-clamp

Plastic work board

Turntable or lazy Susan

Mold release

#3 rebar

Rebar cutter

Container to mix concrete

Trowel

Tools for wet carving (see page 77)

Sharp kitchen knife

Whiskbroom

*If using a premixed material for wet carving, sift the contents of the bag before mixing.

**This should be a mild or galvanized steel. It bends with some effort, but will hold its shape when filled with concrete. Look for 18-gauge metal. Roof flashing will not work for this project.

Garden Torch

DESIGNER: Kem Alexander

Light up your garden with this unique torch. It's a little bit primitive, a little bit Mad Max. Rusted metal found by the artist forms the interesting linear elements, as well as imparts the warm coloration. More than anything else, this project suggests a broader range of possibilities for making concrete projects with seemingly ordinary materials.

Instructions

1 Cut the plastic cone and duct tape it back together keeping the duct tape on the outside of the form. Build a support that will hold your cone in an inverted position. It must be sturdy enough to hold the weight of the metal and concrete and keep the form standing as you work to build the torch.

2 Cut your metal pieces to the desired lengths. Make sure that the metal is clean, dry, and free of any oily residue.

3 Mix your concrete to a thick mud-like consistency.

4 Place about five pieces of metal into the form so they are as flush to the sides as possible. Slowly begin to add your concrete, about a cup at time. As you add more concrete, tap the outside of the form to help eliminate any air bubbles. Add in more pieces of metal close to the sides as the concrete level rises. Continue to tap the form as you fill the cone to the top.

5 Keep building, adding more metal pieces and concrete until you reach the top edge of the cone. Let the form sit undisturbed until the concrete is firm, about 5 hours before carving out the center. Then let the cones continue to set over night.

6 After the concrete has set up firm, remove the duct tape and the plastic cone. Use a small rasp to remove any concrete that has covered the metal.

7 Clean the torch with muriatic acid and water, and then dip it into a solution of water and baking soda to neutralize the effects of the acid. Wrap it in wet towels and plastic and allow it to cure for 30 days. After the piece is thoroughly dried, burn off the fibers and place it in the welded stand.

Cut List for the Mold

Code	Description	Qty.	Materials	Dimensions
A	Legs	4	#5 Rebar	6 feet (1.8 m)
B	Cone ring	1	⅛ x 1½-inch flat metal* found metal ring**	The circumference of the cone measured 3 inches (7.6 cm) down from the cone's top
C	Support ring	1	⅛ x 1½-inch found metal ring**	At least 6 inches (15.2 cm) larger in circumference than the widest part of the cone

*The designer recommends that these pieces be welded together with the rebar on the outside of the rings to provide sufficient support for the concrete cone.

**If you use a found ring, you will need to adjust the size of the cone accordingly.

Materials and Tools

Concrete Mix: Mix 5

Plastic cone

Duct tape

Supports for inverted cone

Selected metal pieces

Container for mixing concrete

Small rasp

#5 Rebar

Muriatic acid

Baking soda

Solar Panel

DESIGNER: **Sherri Hunter**

These panels won't provide heat, but they may bring a warm smile to your face. Using a poured rubber mold is one way to create multiple images. It allows you to develop a mold from a modeled form, providing the artist in you with a new and different outlet for working with concrete.

Materials and Tools

- Concrete Mix: Premixed Sand/Topping Mix, Mortar Mix, Commercial-Grade Masons Mix, Mix 3, Mix 4, or Mix 5
- Plasticine/modeling clay
- ½-inch (1.3 cm) plywood, enough to cut two 13-inch (33 cm) squares.
- Work board (for the bas-relief base)
- Rolling pin or piece of pipe
- Assorted clay tools (optional)
- Work board covered in plastic (for casting)
- Spray acrylic
- Commercial mold release
- Wood for containment frame
- Drill and bit
- 8 drywall screws, 1¼ inches (3.2 cm) long
- Phillips-head screwdriver
- Petroleum jelly
- 2-part urethane mold system*
- 16-oz (.5 L) plastic cups
- Disposable mixing container
- Stirring sticks
- Latex gloves
- Container to mix concrete
- Concrete pigments (optional)
- 14 inches (35.6 cm) of 14-gauge wire
- Needle-nose pliers
- Hammer
- Trowel

* For this project, I used 1 gallon (3.8 L) each of the two components.

Instructions

1 Cut the plywood for the work board for your bas-relief as indicated in the cut list on page 166, and attach them together to make a 1-inch-thick (2.5 cm) base. Roll out the plasticine to a thickness of about ¼ inch (.6 cm), and cover the top and sides of your base board. Design your image and begin to model, adding and subtracting as needed. Use the clay tools to help define your image and add surface textures. When you have finished your model, seal it using the acrylic spray.

2 Cut the wood for your containment frame. Lap one end over the other. Pre-drill two holes at each end to avoid splitting them when you attach them with drywall screws. Use the petroleum jelly on the wood as the mold release. The space between your bas-relief and your containment frame becomes the wall of your finished mold.

3 Refer to the instructions for Flexible Mold, pages 53 to 54, steps 1 through 8.

4 After your mold has set, remove your containment frame. Then, put it back together and set it aside till later. Pull your mold off your model. You should have an exact likeness down to the finest detail.

Working With the System

Read all the instructions that come with your mold-making system. The one I used for this project requires equal parts of two components. I used plastic cups to measure out the quantities needed for each of the components before pouring them into a larger, disposable mixing container.

The mold material will set because of a chemical reaction. That's why stirring for the required amount of time (and then some) is so important. I was a little short of material with my first pour and had to mix more of the mold components. If the same thing happens to you, don't panic and don't rush the steps; measure carefully and stir thoroughly. I had no problem or visible seam line. Make sure your work area is covered and you are wearing latex gloves while mixing and making the mold or things can get very sticky.

Cut List for the Frame and Work Board

Code	Description	Qty.	Material	Dimensions
A	Frame sides	4	1 x 4 stock	4 x 16 inches (10.2 x 40.6 cm)
B	Work board	2	½-inch (1.3 cm) plywood	13 x 13 inches (33 x 33 cm)

5 Clean your mold, the frame, and work area to get ready for casting the concrete. Place the mold face up on the work surface and mark which side is the top. Slide the containment frame back over the mold. This will eliminate any distortion once the mold is filled with concrete. Spray the mold with commercial mold release as directed by the system's manufacturer.

6 Mix your concrete to a soft-batter consistency. Begin applying the concrete in small handfuls, tamping it into the low detailed areas. Continue to fill the mold this way. Use a hammer to tap the sides of the mold to remove air bubbles (I even tap the underside of the table the mold is sitting on). Trowel the top surface.

7 To make the hanger, bend the piece of wire in half using the needle-nose pliers. Bend this loop 1 inch (2.5 cm) from the top. Take the remaining end of wire and bend each side until they loop around like butterfly wings. Find the center point of the top of your casting and push the hanger into the concrete about 4 inches (10.2 cm) from the top. Touch up with the trowel or add a pinch of concrete if necessary. Let the casting sit until hard, 1 to 2 days, before removing it from the mold. Clean the mold, and you're ready to cast another!

Spike Sculpture

DESIGNER: Andrew Goss

Andrew shares several exciting techniques in the construction of this abstract sculpture. While it has a weighty appearance, it's comparatively lightweight because of the armature system. The surface is especially intriguing with its inlaid color and smooth finish. This project will lead you to think of ways you can develop your own vocabulary of sculptural forms.

Materials and Tools

- Concrete Mix: Mix 18 and Mix 19
- Slurry Mix: Mix 21
- Corrugated plastic board
- Plastic packing tape
- Scissors
- Metal mash
- Aviator shears
- Wire cutters
- Container to mix concrete
- Rough sandpaper (16 to36 grit)
- Coarse files and rasps
- Kitchen knives or gouges
- Pigment for concrete (optional)
- Wet sandpaper 100-, 200-, 320-, 400, and 600-grit
- Sealer
- Wax

Instructions

1 To get started on your sculpture, first use Mix 18 for your concrete and follow the instructions for Other Core Materials, pages 61 to 62, steps 1 through 4.

2 The next day, check to make sure that your concrete is hard enough to not break, but soft enough to sand. Use the coarse sandpaper or file to refine the shape of your sculpture. Concentrate on the overall form rather than the surface texture. Wash the concrete to remove any dust, but blot off extra water from the surface.

3 Mix the smooth concrete, Mix 19, and refer to the instructions for Inlaid Color, page 113, steps 1 through 5, to finish your piece.

Wacky Totem

DESIGNER: **Sherri Hunter**

You might ask, "What was she thinking?" when you see this project. Maybe it's a little about the 1960s and a lot about seeing how many different processes I could use in one piece. That's one of the reasons I love totems. Ideas, skills, and techniques can each be explored in small scale, and then combined to make a much bigger statement. You can choose specific themes like the totems pictured on pages 18 and 21—or just be wacky!

Materials and Tools

- Concrete Mix: *Boulder* Premixed Mortar Mix, Commercial-Grade Mason Mix, or Mix 3; *Bowling Pin and Sphere* Mix 12; *Flower Topper* Premixed Mortar Mix, Commercial-Grade Masons Mix, or Mix 3
- Slurry Mix: *Bowling Pin and Sphere* Mix 21
- All the Materials and Tools listed for the Faux Boulder, page 133
- PVC pipe, with 1⅜-inch (3.4 cm) inside diameter*
- 20-gauge galvanized or stainless steel wire
- Polystyrene
- Polystyrene carving tools, page 56
- Light weight AR mesh or tape
- Container for mixing concrete
- Container for mixing slurry
- Paintbrush
- Assorted mason stains
- Table knife
- Mirror
- Tile cutter or glass cutter
- Diamond sharpening block (optional)
- Thin-set adhesive
- Grout
- Found truck spring
- Steel wool
- Rust-inhibiting paint
- Found porcelain escutcheon
- Polystyrene insulation board
- Roof flashing
- Assorted bottles with lids or corks
- Sand
- Spoon
- Misting bottle
- Reinforcing rods
- Sea glass and glass gems
- Sponge
- Heavy pipe for installation, with 1½-inch (3.8 cm) outside diameter*

*It's important to check the diameters of your pipes before you begin your construction. You want the PVC pipe that you use in your elements to be the same size. Check that the PVC will slide smoothly and snuggly over your installation pipe. You also don't want to have an extra gap that will allow the pieces to rock on the pole when they are installed.

Instructions

1 Follow the instructions for making a Faux Boulder, pages 135 to 136, steps 1 through 9. However, before you start to apply the concrete, you need to cut a piece of PVC pipe 2 inches (5 cm) longer than the height of the middle of the boulder armature. Cut a hole in the expanded mesh, insert the pipe and secure it to the armature with additional wire, taking care that it is positioned perpendicular to the ground.

2 Carve the polystyrene to your desired shape (in this project I used a bowling-pin form), referring to the instructions on Polystyrene Foam Armatures, pages 55 to 56. When you are happy with the form, follow the instructions for Adapting a Polystyrene Foam Core for a Totem Construction, page 59, steps 1 through 9.

3 Prepare the sphere using all the instructions referenced in step 2.

4 Apply concrete to the two polystyrene forms using the instructions for Polymer Fortified Concrete and Alkali Resistant Fiberglass Mesh, pages 82 to 83, steps 1 through 6.

5 The surface texture on the bowling pin was achieved by applying multiple coats of colored, fortified concrete, with a table knife. The concrete was mixed one color at a time and applied in thick random strokes.

NOTE Wipe the knife blade clean during the application of each color to prevent the colors from becoming muddy.

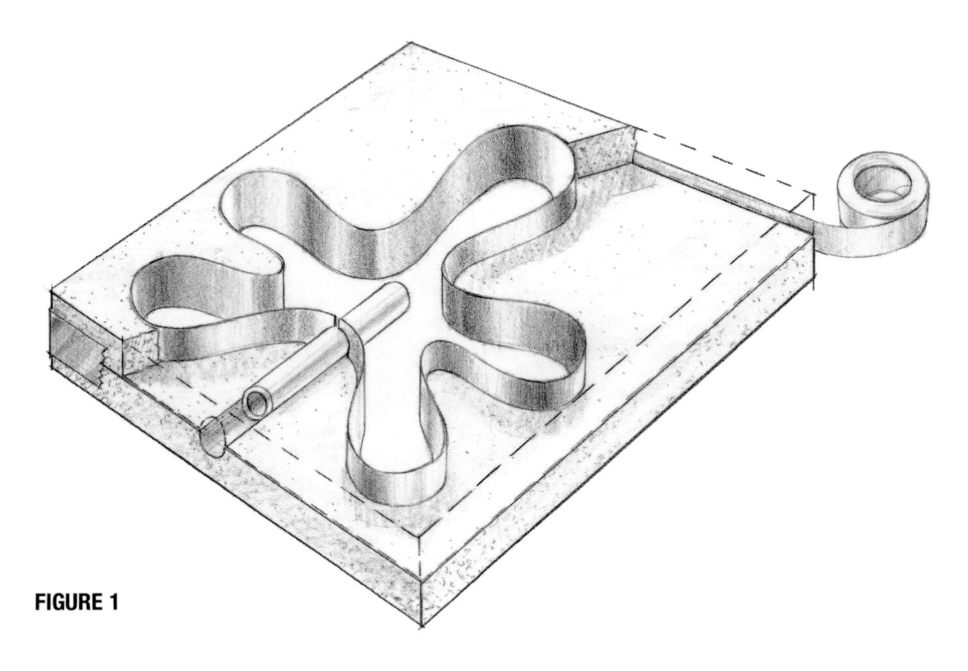

FIGURE 1

6 To mosaic the sphere with mirror, refer to pages 99 to 100. Cut a good-quality, commercial-grade mirror into squares. Use the diamond-sharpening block to slightly bevel the sharp edges. Find the middle of your sphere by measuring the distance between the top and bottom pipe. It helps to get your lines level by starting at the middle and working up towards the pipe. Turn the sphere over to complete the mosaic. The mosaic process will proceed much faster if you use one of the stands described in the sidebar. After the mosaic has dried overnight, grout it by referring to the instructions on pages 101 to 102.

7 Found objects add to the wackiness of this totem. I cleaned a truck spring with steel wool and then painted it with a rust-inhibiting paint. An old porcelain escutcheon worked like a washer to keep the top piece from touching the spring.

NOTE Improvise! There are lots of interesting found objects waiting to be discovered.

8 I have to admit to a little influence from Ricky Boscarino here. This piece of the totem will combine Polystyrene Foam Waste Mold, page 43, with the Bottle Panel Project on pages 157 to 159, and the Flower Power project on page 142. To insert the pipe, review Adapting a Polystyrene Foam Core as a Pole Topper, page 60, with the understanding that this pole will be cast into the piece. Refer to figure 1 to see how these different techniques have been combined.

9 Cut lengths of thin rod and bend them to extend into the petal extensions. Take into consideration that they can't cross over the bottles. Set aside until you are casting. Once your bottles are positioned, mist the sand with water and wipe the bottles dry.

10 Mix your concrete to a soft-batter consistency. Start by filling in the concrete around the bottle necks, and then continue to fill your mold until the base of the bottles are almost covered.

11 Let the concrete set until firm enough to embed any additional decoration. When the concrete is firm enough, carve out around the bottles and smooth with a sponge. Let the form remain in the mold for 2 to 3 days. Carefully break away the mold, remove the metal and lift the piece out of the sand. Take the piece outside and scrub off the remaining sand.

12 The stem of the flower was covered after the piece was out of the mold. Wrap the piece of pipe with AR mesh and paint it with a concrete bonding agent before applying the concrete and embedding it with glass gems.

13 After all the pieces have cured, you can begin your installation. Find a level spot for the boulder base. Insert the installation pipe and begin to thread the pieces over the pipe. You may want to put large rubber washers between each piece.

NOTE To determine the length of your installation pipe, measure each piece, including the top piece, by inserting a tape measure into its PVC pipe. Total the measurements and subtract 2 inches (5 cm). If you're adding washers, include their measurements as well. For a more permanent installation, dig a hole and cast a section of the PVC pipe in place with concrete, making sure that the pipe is plumb. Add the length of the foundation pipe when computing the length of your installation pipe.

Work Stand

To work on individual pieces, I build a simple stand using a 12-inch (30.5 cm) piece of 2 x 12 stock lumber. This allows me to work without having to lay the pieces down. To make a stand, find the middle of the board and use screws to attach a metal plumbing flange to it. Then screw a length of threaded pipe into the flange.

Three Arches

DESIGNER: Lynn Olson

When you first look at one of Lynn Olsen's sculptures, you're captivated by the polished surface—it looks like marble, but it's not. This elegant sculpture utilizes the ferro-cement construction and polishing techniques developed and refined by Lynn. His processes can be adapted to create figures, animals, or abstract structures like this one.

Materials and Tools

- Concrete Mix: Mix 14 or Mix 16, and Mix 12 (Sand-Cement Mix) and Mix 17
- Slurry Mix: Mix 21
- Needle-nose pliers
- Angle wire cutters
- ¼ and ½-inch (.6 and 1.3 cm) stainless steel rods
- 18- or 20-gauge stainless steel wire
- Work board covered with plastic
- Container for mixing slurry
- Paintbrush
- Container for mixing concrete
- Concrete pigments (optional)
- Sharp kitchen knife with ½ inch (1.3 cm) of the tip bent at a right angle
- 2 ml plastic sheeting
- Assorted coarse rasps
- Assorted files
- Riffler rasps and files
- Wire brush to clean rasps and files
- Silicon carbide paper in various grades from coarse to very fine (600-grit)
- Methyl methacrylate/acrylic polymer (optional)
- Clean cotton gloves (optional)
- Wax (optional)

Instructions

1 Review the information on Metal Armatures, page 62.

2 Design your basic form using sketches or a maquette. Refer to the instructions for Ferro-Cement Animal, pages 72 to 73, steps 1 through 7. If you're working on an abstract piece, like Three Arches, just substitute the word sculpture for animal.

3 You'll need to work with the set stages of your concrete as you develop your sculpture. After the final coat of fiber cement has set for 24 hours, you can refine the shape with rasps, files and even the coarsest silicon carbide abrasive paper. Remember to continue to keep your piece damp while you're working on it to avoid dust and to facilitate the hydration process. As the piece continues to harden, use finer files and finer grades of silicon carbide paper to develop a smoother finish. Use the final 600-grit silicon carbide paper to produce the final polished surface after the concrete has cured for about 4 weeks.

4 Applying a sealer to the finished piece will help prevent stains as well as add gloss to the surface. A water-based methyl methacrylate (acrylic polymer) is recommended. This sealer will inhibit moisture from penetrating the surface while still allowing the vapors to escape. The best way to apply it is to wear gloves to rub it on by hand until it dries. Buff the surface with a soft cloth to achieve a luster and to enhance the colors and grain patterns. If you want more gloss, use a wax that contains carnauba.

Little Girl

DESIGNER: Sherri Hunter

You might not know her, but she'll still extend a welcoming smile to all as she presents her latest garden find. This young lady was formed over a metal armature and has directly modeled and carved details. You can adapt this same process for sculptures and structures of all descriptions as seen in "Make Your Dreams Concrete!" on pages 115 to 117. The most important thing to remember when making any modeled project is to just let your imagination run free.

Instructions

1 Follow the instructions for Simple Figurative Armature, pages 69 to 72, steps 1 through 10.

2 The purpose of the mesh is to define the form. When applying the mesh, work from the inside out. For example, cover the legs and the body before adding the skirt. You want to add sections of mesh to help complete the gesture of the figure without building up too much bulk. Using stiff wire along with the mesh will also help. Look at shapes around you when trying to figure out how to cut your mesh. The skirt on this figure is shaped more like a lampshade; the circumference at the top (around the waist) is smaller than the circumference at the bottom (the hem). The hem edge was reinforced by folding up 3 inches (7.6 cm) of mesh. In order to make the mesh fold up and lay flat, cut slits in the mesh as you fold.

3 Follow the instruction for Modeling, pages 74 to 76, steps 1 through 6.

NOTE Sometimes when I've been modeling for a while and the concrete has firmed, I'll decide to carve away some of the sections to either define a shape or add texture. As you bring your figure to life, think in terms of using both additive and subtractive processes.

Cut List for the Base Form

Code	Description	Qty.	Material	Lengths
A	Sides	2	2 x 4 stock	3½ x 14 inches (8.9 x 35.6 cm)
B	Ends	2	2 x 4 stock	3½ x 27 inches (8.9 x 68.6 cm)

Cut List for the Armature

Code	Description	Qty.	Material	Lengths
A	Base head/leg support	2	#3 rebar	50 inches (127 cm)
B	Outer head/leg support	2	#3 rebar	45 inches (111.8 cm)
C	Outside leg	2	#3 rebar	22 inches (55.9 cm)
D	Hip	2	#3 rebar	12 inches (30.5 cm)
E	Hip cross pieces	2	#3 rebar	6 inches (15.2 cm)
F	Waist	2	#3 rebar	8 inches (20.3 cm)
G	Shoulder	2	#3 rebar	12 inches (30.5 cm)

Materials and Tools

- Concrete Mix: (Base) Premixed Concrete, Fast-Setting Concrete, or Mix 1, (Figure) Premixed Mortar Mix, Commercial-Grade Masons Mix, Mix 3, or Mix 5
- Slurry Mix: Mix 20 or Mix 21
- 2 x 4 wood for building a base form
- Drill
- 8 drywall screws, 3 inches (7.6 cm) long
- Mold release
- Container for mixing concrete
- 4 ml plastic sheeting
- Screed
- Trowel
- Pointing trowel
- Margin trowel
- Leather work gloves
- Heavy-duty metal cutter
- #3 Rebar
- Rebar bender
- Grinder or metal file
- Lineman pliers
- Twisted tie wires (or roll of tie wire cut to comfortable work lengths)
- Wire twister/winder
- Expanded metal mesh
- Hog ring pliers (optional)
- Hog rings (optional)
- Rub block
- File
- Container for mixing slurry
- Paintbrush
- Spray bottle
- Sponge
- Short piece of ¼-inch (6 mm) dowel, sharpened in a pencil sharpener
- Assorted artist brushes
- Kitchen knife
- Plastic drop cloth
- Assorted wooden or wire-end clay or sculpture tools (optional)

Acknowledgments

I want to thank Lark Books for giving me the opportunity to write another book. Thank you to Jane LaFerla, my editor, for your patience and persistence. Thanks also to art director Stacey Budge for her transformative gardening skills and her superb talent in designing this beautiful book. Thanks also to art directors Chris Bryant for the how-to photography and Kathy Holmes for filling in as needed. To photographers Stewart O'Shields, Evan Bracken, and John Widman, thanks for adding your talents to the book, and to Olivier Rollin for his illustrations.

I wish the beautiful gardens in the photos were mine, but they belong to several generous Asheville, North Carolina, residents including Terry Taylor, Chris Bryant, Susan McBride, and The White Gate Inn, a local bed and breakfast.

Much thanks and credit goes to the other artists for their designs and technical information: Kem Alexander, Ricky Boscarino, Virginia Bullman, Andrew Goss, Marsha Hoge, Elder Jones, Lynn Olson, Tom Rice, Dixie Stephens, and Marvin and Lilli Ann Rosenberg. Thanks also to all the artists who took the time to contribute Gallery images for the book. A special thanks is extended to Willem Volkersz, for sharing the images used for Make Your Dreams Concrete.

Thanks to my studio assistants: Kristina Bell, Janet Cataldo, Levi Cataldo, Tiffany Delk, and Harry C. Kellogg. And through it all, I'm very thankful for my husband Martin. His love, friendship, patience, and assistance help to make good things possible, even moving boulders. What a team! Thank you all...I couldn't have done it without you.

Contributing Artists

Kem Alexander
301 E. Knoll St., Chattanooga, TN 37405
(423) 505-1227
www.kemalexander.com

Ricky Boscarino
22 DeGroat Road
Montague, NJ 07827
(973) 948-2160
www.lunaparc.com

Virginia Bullman
Virginia's Art Etcetera
2435 Tall Pines Lane,
Hillsborough, NC 27278
(919) 933-4950

LaNelle Davis
Contact through Virginia's Art, Etcetera (see Virginia Bullman)

Faducci
P.O. Box 923
North San Juan, CA 95960
(530) 292-3857
www.faducci.com

Andrew Goss
718 Second Ave. West
Owen Sound, Ontario, N4K 4M2, Canada
(519) 371-1857
www.markersgallery.com/concrete

Johan Hagman
6123 Stonehaven Dr.,
Nashville, TN 37215
(615) 661-8854;
www.artsnashville.org/registry/artsitaz.html

Kathy Hopwood
409 Woodland Park Dr.,
Hillsborough, NC 27278
(919) 644-1335
home.earthlink.net/~kathyhopwood

Sherri Warner Hunter
3375 Fairfield Pike
Bell Buckle, TN 37020
(931) 389-9649
www.sherriwarnerhunter.com

Elder Jones
P.O. Box 81
Readyville, TN 37149
(615) 409-6005;
www.sandpudding.com

K.C. Linn
420 No. Spring St.,
Murfreesboro, TN 37130
(615) 890-2006

Lynn Olson
4607 Claussen Lane
Valparaiso, IN 46383
(219) 464-1792;
Call for information on his book Sculpting in Cement

Tom Rice
4715 Peytonsville, Rd.
Franklin, TN 37064
(615) 595-7113

Marvin and Lilli Ann Killen Rosenberg
4001 Little Applegate Road,
Jacksonville, Oregon 97530
(541) 899-7861

Phil Schuster
410 North Albany St., Chicago, IL 60612;
(773) 265-1848

Index